PREFACE

It is appropriate that the Society of Automotive Engineers review its contributions to society during the Diamond Jubilee Celebration. The evolution in farm mechanization over the last 75 years has been a major factor in enabling the agricultural segment of our economy to be a leader in increased productivity. Ingenuity and the quest for profit have resulted in equipment designed for greater efficiency, increased capacity, and improved comfort and safety.

During the last 75 years tractors have replaced animals as the power source on the nation's farms. Coleman and Burnham review the milestones in farm machinery evolution. During this evolution, cumbersome, low productivity machinery has been replaced with easily operated, improved high performance equipment incorporating many safety and comfort features.

As tractors were introduced to the farm scene, there were reliability problems with both machines and manufacturers. Candee and Walters describe the history of comparative tractor testing under uniform conditions. A voluntary test code was developed which has worldwide recognition. Documentation is given on revisions made to keep the test code current with changing requirements.

Pekar reminds us that society has recognized and encouraged innovative efforts through the patent system which has served agriculture since early colonial days. Changes have been made in the system to meet the needs of changing technology and the economy. This is a vital service which combines legal and technical procedures for inventors and is beneficial to everyone.

Murray affirms that exclusive rights (patent powers) as provided by Congress, promote the progress of science and useful arts. Engineers are prone to overlook patentable ideas because, with hindsight, they rationalize that *anyone could have done it.* The engineer and his patent attorney must deal with questions of infringement, usefulness, and obviousness.

Each step in the evolution of labor saving agricultural machines has eliminated some existing hazards but created new ones to challenge the designer. Zink reviews these problems and the progress made in safety engineering with the introduction of complex machines. The farm machine industry has diligently created, and encouraged conformance to, voluntary standards which incorporate recommended safety practices.

The power take-off drive has provided an economical and versatile power source for machines. Morrell outlines stages of development, performance, and hitching problems. Voluntary industry standards have been written and updated to keep abreast of changing usage patterns and increasing size of equipment. These standards make it possible to have interchangeability of tractors and machines without an array of adapter bundles which would be an irritation to the user, increase exposure to hazards, and increase the cost of operation.

This is an interesting historical review of 75 years of progress in farm mechanization. Credit is given to the ingenious farmer-inventor and to those in the industry who materialize these ideas. There were challenges in the past to provide safe, efficient, highly productive equipment. We face these same challenges in the future with an added emphasis on conservation of our resources.

Wendell M. Van Syoc

An Historical Perspective of Farm Machinery

SP-470

Published by:
Society of Automotive Engineers, Inc.
400 Commonwealth Drive
Warrendale, PA 15096
September 1980

ISBN 0-89883-241-1
SAE/SP-80/470

TABLE OF CONTENTS

800930

History of Society of Automotive Engineers Agricultural Tractor Test Code

Russell Candee and F. C. Walters
John Deere Product Engineering Center
Waterloo, IA

We should review some of the historical background before discussing the history of the Society of Automotive Engineers (SAE) Agricultural Tractor Test Code.

Mr. R. B. Gray, in his compilation of information entitled "Development of the Agricultural Tractor in the United States,"[1] credits the invention of the "steam plow" or "traction engine," circa 1855-1858, with being the first step of importance in mechanical power farming in this and other countries. In addition to the demand for increased power for plowing, various new machines invented in the last quarter of the nineteenth century stimulated a need for more mechanical power than that which could be supplied by animals. "Plowing tests" or "plowing and threshing exhibitions" were a common events during the last quarter of the nineteenth and early part of the twentieth century.

Despite all this activity, it was not until 1908, during the Winnipeg, Canada, Canadian Industrial Exhibition, that public "agricultural motor competition" (tests) were held. These trials were conducted for both belt pulley and drawbar (generally plowing) work. The "rules" or test procedure during these trials,[2] 1908-1913, consisted basically of the following:

1. Belt pulley work with the traction engine belted to a Prony brake dynamometer. Figure No. 3
2. Drawbar - plowing field test - with a pull-recorder between plow and tractor. Figures Nos. 1 and 2.

Abstract

This paper outlines some of the history involved in the development of the SAE/ASAE Agricultural Tractor Test Code, along with some thoughts for the future. Much credit should be given to the men of the tractor industry, the Nebraska tractor test group and the technical societies for their vision and leadership. This work has enabled us to arrive at the position of creditability with respect to tractor power performance that we have with our customers. The integrity and dedication we have within our tractor industry, which includes all phases; namely, the companies, the technical societies and the Nebraska Tractor Test Laboratory, the writers feel confident we will continue to meet and solve the problems arising in the future.

SAE/SP-80/470/$02.50

Figure 1
Top view of Schaffer and Budenberg drawbar pull meter used at the Winnipeg, Canadian Industrial Exhibition "Agricultural Motor Competition" (1908-1913)

Figure 2
Side view of drawbar pull meter

Figure 3
Type of Prony brake (thought to be the original Prony brake) used to determine belt pulley power during the Winnipeg, Canadian Industrial Exhibition (1908-1913)

Figure 4
Tractors belted to the Prony brake during the Canadian Industrial Exhibition (1908-1913)

The drawbar tests were basically plowing, but the drawbar pull, fuel consumption, drawbar horsepower or belt horsepower hours per unit of fuel were determined. A scoring system was established on which prizes were awarded.[3] Although probably not envisioned then, these draft rules might be credited as the basis for tractor test codes later.

The Winnipeg agricultural motor competitions (tests) were followed by similar "national demonstrations" held at:

1. Fremont, Nebraska (1911-1918)
2. Salina and Wichita, Kansas (1918-1920)
3. Atlanta, Georgia (1918-1919)
4. Fargo, North Dakota (1918-1919)
5. Ohio (sites selected by Ohio State University) (1919)

All these demonstrations generally followed the guidelines or rules used at the Winnipeg exhibition (1b, 1c, 3).

A "test code" or "guidelines" for scoring the participants is as recorded and published by L. W. Chase in 1917. It is of interest to note that Professor Chase, Chairman of the Agricultural Engineering Department of the University of Nebraska, had written his masters thesis on the subject of testing agricultural tractors.[2] At this time field demonstrations were extremely popular, with demonstrations being carried out by manufacturers, dealers, county agents,

--often each making his own rules or guidelines. It might be surprising to some, but there was a Nebraska Tractor Test procedure established in 1917 by Professor Chase.[3] This 1917 procedure when combined with the 1917 SAE engine test code and the SAE/ASAE tractor rating code, has similarity to the Nebraska tractor test "rules"[8] later established in 1919 and still, in modified form, used today. The shortcoming of these demonstrations was the difficulty for the farmer in translating the results into terms of work on his farm, plus the fact that great amount of field space was required for the plowing tests.

The period before and after World War I (1912-1922), at best, might be characterized as one having exaggerated power performance claims in the agricultural tractor field. Many of the tractors had poor mechanical reliability; companies often had a poor warranty policy and poor to nonavailability of repair parts service.[1d6a1f] During this period, the number of tractor companies varied from 31 to 166. Many of these tractor companies were buying engines, transmissions, chassis, etc., assembling such, and offering the unit for sale, using a variety of sales tactics (including the sale of stock in the company). At the same time, farmers were changing from animal power to tractor power and had little or no experience as to what might be rightfully expected performance from a "tractor".

The situation reached a point where, crica 1915, the farm equipment industry and others agitated for standardized ratings for tractors, to the extent that a bill was introduced to the Congress of the United States, at the request of the American Society of Agricultural Engineers (ASAE), favoring the creation of a separate commission under the U.S. Department of Agriculture, part of whose job would be the testing of tractors. The commission was to be known as the Bureau of Agricultural Engineering.[13, 6b] After considerable discussion, no appropriations resulted; consequently, the endeavor failed.[1d, 6a]

During this same period both the American Society of Agricultural Engineers (ASAE) and the Society of Tractor Engineers of Minneapolis were discussing the need for some type of tractor rating. The Society of Tractor Engineers (later to be merged with the Society of Automobile Engineers in 1916, with the combined group still later changing its name to Society of Automotive Engineers) held a meeting during the 1917 Fremont tractor demonstrations.[4b] Several problems requiring standards were discussed during this meeting. One of the needs was a "Standard Rating for Tractors" including the necessary tests. This need resulted in the SAE (August, 1917) and later [1d] the ASAE adopting an Agricultural Tractor Test and Rating Code.[5] This code basically consisted of two parts:

1. A Belt Power Test and Rating: The code reading, "The belt power rating shall be 80 percent of the horsepower the engine is guaranteed to deliver at the belt pulley continuously for two hours, the engine being in good condition and properly operated at rated speed".
2. The Drawbar Test and Rating: The code reading, "The drawbar power rating shall be 80 percent of the horsepower that the tractor is guaranteed to develop at the drawbar continuously for two hours, the tractor being in good condition and properly operated at rated engine speed. The test should be taken on ground sufficiently firm to give the tractor wheels a good footing, a firm sod being preferable".

The SAE and ASAE code for tractor ratings provided a procedural code whereby agricultural tractors could be rated as was commonly being done by various organizations and individuals. This is a good example of industry and technical individuals forming a "voluntary standard" that has contributed greatly to the agricultural industry with little burden.

As a result of the failure of any national action on this situation, four states; namely, Nebraska, North Dakota, Illinois and Wisconsin had bills introduced into their 1919 legislative sessions requiring some type of tractor testing.[6b] Other states, for example, Indiana and Missouri had the subject under discussion, but it appears that such discussions did not reach the bill stage. The House Bill No. 85, introduced by Representative W. F. Crozier, in the Nebraska Legislature (16 January 1919) and signed by the governor on 20 March 1919,[7] was the first to be enacted into law on this subject. The law stated that "On and after July 15, 1919, no tractor or traction company shall be permitted to sell or dispose of any model or type of gas, gasoline, kerosene distillate or other liquid fuel tractor engine in the State of Nebraska, without first having said model tested and passed upon by a board of three competent engineers who are or shall be under the control of the State University management. Each and every tractor presented to the State University management for testing shall be a stock model and shall not be equipped with any special appliance or apparatus not regularly supplied to the trade. A tractor not complying with the provisions of this section shall not be tested under this act, nor the result certified. Sec. 2. Test, of what to consist. Such tests to consist of endurance, official rating of horsepower for continuous load, and consumption of fuel per hour or per acre of farm operations; the results of such tests to be open at all times to public inspection" -- plus other administrative features.[7] Administration of this law was assigned to the Railway Commission, which in turn gave the task to Professor L. W. Chase, Chairman of the Agricultural Engineering Department of the University of Nebraska, for establishing "the program of organizing and conducting the testing procedure". The test procedure or "Rules" that resulted is recorded in Agricultural Experiment Station - The University of Nebraska, Lincoln, Bulletin No. 10 -December, 1919. The test consisted of nine basic parts as follows:

1. Limbering up run
2. Brake horsepower test at rated load
3. Brake horsepower test at varying load
4. Brake horsepower test at maximum load
5. Brake horsepower test at half load
6. Drawbar horsepower test at rated load
7. Drawbar horsepower test at maximum load
8. Miscellaneous
9. Endurance

Activities at the University of Nebraska the year following passage of the law verified the situation that existed in the tractor industry. During the year following passage of the law, the University of Nebraksa Tractor Testing Laboratory:

1. Reviewed 100 applications for tests of particular models of tractors.
2. Received 68 tractors.
3. Completed tests on 65 tractors of which 36[12a] had to be repaired during the course of the test.
4. Required four traction (four tractor) companies to materially revise their advertising claims prior to being granted certifications.
5. Accepted withdrawal of three tractors after the review of their test performance results.

Although the Minneapolis (listed as Twin City) 12-20[12a] tractor was the first tractor to start the test (fall of 1919) which were interrupted by snow, the John Deere Waterloo Boy, Model N, tractor was the first to complete the test program on 9 April 1920, a little over a year after the "Tractor Test Bill" was signed.

Considering the development of the Agricultural Tractor Test Code, one must also review the test procedure or "rules" used by what later became known as the Nebraska Tractor Test Laboratory. Furthermore, since the SAE and ASAE and the Nebraska group are so closely involved, it would be improper to attempt to separate or try to identify in detail contributions made by each of the groups.

Table No. 1 shows a comparison of a brief of the contents of the SAE/ASAE Test Code and the Nebraska Tractor Test "Official Rules" that were in use during the different periods. Although the code and "Rules" still maintain the basic fundamentals identified in the pre-1920 era; namely:

1. To measure the tractor's power for operating implements at the power outlets (belt pulley or power take-off and drawbar)
2. To communicate this information to the public

there have been many evolutionary changes following the introduction of the first SAE/ASAE-Nebraska Test Code (1917). A few of these changes are as follows:

1. Testing the tractors on a test course instead of plowing in field (1920).
   a. The first test was done on a half-mile, hard dirt test course with both level and sloping portions.
   b. This was followed by changing the surface to
      1. A 1000 ft. cinder-clay course*[8b] Figure No. 5.
      2. Later, to an oval clay test course whose two straight portions are level and with the straight portion of sufficient length for a 500 ft.** measured test run.
      3. Finally (1956) to an oval concrete test course with the same length dimension as (2) above.

*From interviews with Mr. L. F. Larsen and Mr. C. L. Zink, formerly chief engineers for the test, this cinder course may have been used only a short time, followed by the return to the clay test course.

**At one time this measured distance was 250 ft., but was later change to 500 ft.

Figure 5
Tractor testing course, University of Nebraska circa 1920 (ref. SAE transactions 1920, Part I, page 640)

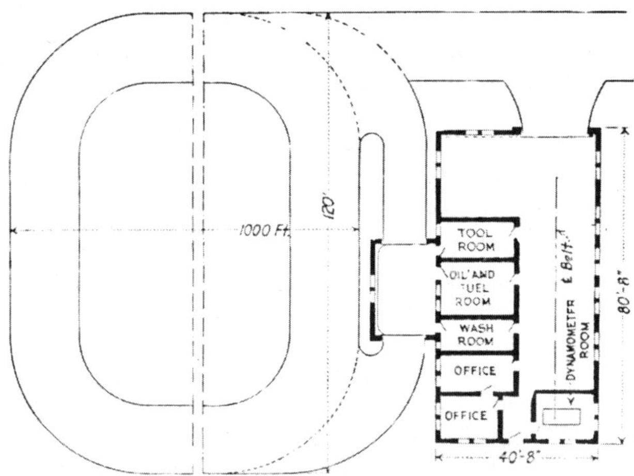

2. Providing at certain periods of time, various SAE-ASAE tractor power ratings whose purpose was to provide the buyer some idea of a means for matching his tractor and implements. The SAE-ASAE tractor power rating system was abandoned in later years.
3. Providing a system for determining and reporting the tractor's performance in terms of "corrected horsepower" which has been abandoned in later years (1959).
4. Recognizing the difference in operating conditions between steel wheels and rubber tired tractors by incorporating (circa 1941) the drawbar testing of both a ballasted and unballasted tractor; later as the use of rubber tires became standard and the characteristics of rubber tire tractors became well-known, the test at the unballasted tractor weight condition was abandoned (1972).
5. Recognizing the change in trend from belt pulley to power take-off operated implements, by incorporating tests at rated PTO speeds (circa 1959).
6. Providing a procedure for determining the torque reserve* characteristics of the engine as installed in the tractor (1952); first by measuring the torque at the belt pulley or power take-off and later shown in the "lugging run".

   *When conducting the "Nebraska Tractor Tests", the torque value was calculated and used as a measure of load for the varying power tests. Starting 1956, the lugging test was conducted.
7. Providing a procedure for measuring and reporting the sound levels of the tractor, both at the operator's station and "drive by", as determined under test course conditions (1972).

A review of Tables No. 1 and 3 shows the results of the intensive attempt by all groups; namely, the tractor industry, the technical societies, the personnel of the Nebraska tractor test laboratory (both the Nebraska test Board of Engineers and test personnel, and the Nebraska Legislature) to maintain a practical close agreement between the SAE-ASAE test code, the Nebraska Legislature's Enabling Act, and the Nebraska "Official Rules" for testing tractors. This speaks highly of the objectiveness in attitude, considerations for practical approach and the fine cooperation that has existed between these groups through the last 60-plus years, despite oftentimes the sincere difference on points of view that arise between engineers with different backgrounds, interests and professional environments. This accomplishment is a testament to the "voluntary system" for the forming and development of "standards" as used in the United States.

Having briefly reviewed some of the history behind the SAE-ASAE Tractor Test Code and the Nebraska "Official Rules" for conducting tractor tests, we should look at the results obtained, comparing the situation today with that existing 61 years ago.

1. A means has been provided whereby the power and performance characteristics of the tractor, under controlled conditions, can be factually communicated to the public, governments, other companies and engineers with the acceptance of the information received with a large measure of confidence.
2. A means has been provided through the use of basic data whereby the performance of a given model and/or make of tractor can be compared with other tractors offered for sale in the United States and by use of the nondimensional parameter's technique within practical

limitations can be compared on a worldwide basis.

3. A means has been provided whereby the customer can compare the performance claims made either by the manufacturer's or dealer's personnel with the performance data obtained by the Nebraska Tractor Test Laboratory. This, to a large measure over the past years, has led to correcting the situation that existed during the period of 1912 to early 1920's as described by Representative Crozier[1d, 6a] of the 1919 Nebraska Legislature.

In international trade, communication on a mutually-understood basis is mandatory. Table II shows chronologically the chain of events, as known to the writers, relative to the testing of agricultural tractors in North America and other areas in the world. This Table shows the date when a type of testing of Agricultural tractors to a performance code was started in the different countries. Where two dates are shown, it appears that at the earlier date, the power at the belt pulley (or power shaft engine power) was determined by an engine type dynamometer, while the drawbar pull (and drawbar power) was determined by using a form of pull meter during plowing operations (a method similar to that developed during the Winnipeg Agricultural Motor Competitions). The later date was when testing was started on a test course to determine the drawbar power. A review of the references cited leads one to believe that the same need for comparative information existed at about the same time in countries such as England, Spain, Sweden, Australia, as existed in North America. The writers do not know whether the communication during the era of 1908-1919 was such that one group of Engineering-oriented people were aware of what similar groups in different countries were thinking, but the references again indicated when similar groups are faced with similar requirements, similar solutions and development of solutions result. Again reviewing the references, it appears that the following milestones for quantifying the performance of agricultural tractors originated as follows:

1. The Winnipeg Agricultural Motor Competition introduced the scheme of quantifying agricultural tractor performance by testing the tractor's power at both the belt pulley and drawbar (1908).
2. The Nebraska Tractor Test "Rules" and SAE-ASAE Tractor Test Codes introduced the scheme whereby the tractor's drawbar power performance was quantified by testing on a test course rather than under a plowing operation (1920).
3. The Swedish Tractor Test Code (1956 and slightly later dates the OECD Tractor Test Code) introduced the scheme for quantifying the strength of agricultural roll-over protective structures.
4. The Australian Tractor Test Code (1962) introduced the scheme for quantifying the sound levels of the tractor.

Having briefly reviewed the past and present situation of the SAE-ASAE Agricultural Tractor Test Code, where should we proceed from here? Each individual is entitled to his own point of view on this question; however, the following thoughts are those of the writers:

1. Our goal should be to test tractors per Agricultural Tractor Test Code ISO 789 as it now exists or as revised. The present situation involves testing to various national and technical codes. This duplication results in unnecessary expense to the customer and utilization of resources that could be used more advantageously. Increasing world trade continues to add complexity to the situation.
2. There should be reciprocity in acceptance of test results between test stations and governmental bodies. Testing to a common ISO code to which all interested groups have had opportunity to make input would provide a common basis for worldwide use. Testing to the ISO Agricultural Tractor Test Code ISO 789, using the sections as applicable, could be the basis for reciprocity of acceptance of test results.
3. Some of the new tractors being developed for the future could be of a size which could exceed the capabilities of the present test station facilities. To make better use of preparation time, manpower and money available for test equipment, where the accuracy of the manufacturers test equipment and facilities meet the accuracy criteria established by the test station, it might be also possible to conduct tests at the manufacturers facilities under supervision of test station or other appropriately certified personnel. This would provide a means whereby the number of

units tested per year and the corresponding test results made available, could be increased with only nominal increase in financial appropriations.

4. In view of the many tractor tests throughout the world, it would be advantageous if one agency had a central worldwide file of agricultural tractor test results. This source could list tractors in production on 1 January of the year involved. A performance summary could be issued each year. The Nebraska Tractor Test laboratory issues such a summary on the basis of tractors covered by their tests.
5. New tractor models may differ from previous models in only a few different features or characterstics. For the purpose of economy, it should be possible to test only new or different characteristics rather than repeating the complete test, thereby saving time and money. To accomplish this facet, the Agricultural Tractor Test Code should be broken into individual sections (or individual test codes). This would enable the manufacturer and the testing laboratory personnel to select the appropriate tests without redoing the test of items unchanged on the tractor.
6. In the 1908-1920 period, one of the shortcomings of the demonstrations was the farmer's difficulty in translating the "results of the tests" into terms of work on his farm. Technology exists today whereby application of the information from the test results, information pertinent to the farming operation, could be made available. Educational programs on the application of the test information would be beneficial to both the farmer-buyer and the tractor industry.
7. Consumers may desire more information than is available. The writers are sympathetic to the consumers' desire for more factual information and they believe it is the manufacturer's responsibility to fulfill these needs. However, if the "reviewing agencies" feel the above inadequate, then the additional information should be developed through cooperation between the manufacturer and the "reviewing agency". In the interest of economy of both time and money, we should devise a system whereby we could take advantage of test results from manufacturers' various test programs rather than continuing to retest as we are doing today.

In summary, we have briefly outlined some of the history involved in the development of the SAE/ASAE Agricultural Tractor Test Code, along with some thoughts for the future. Much credit should be given to the men of the tractor industry, the Nebraska tractor test group and the technical societies for their vision and leadership. This has enabled us to arrive today at the position of creditability with respect to tractor power performance that we have with our customers. With the integrity and dedication that we have within our tractor industry, which includes all phases; namely, the companies, the technical societies and the Nebraska Tractor Test Laboratory, the writers feel confident we will continue to meet and solve the problems arising in the future.

Figure 6
Plan and elevation of testing car (circa 1920. Ref. SAE transactions 1920, Part I, page 641)

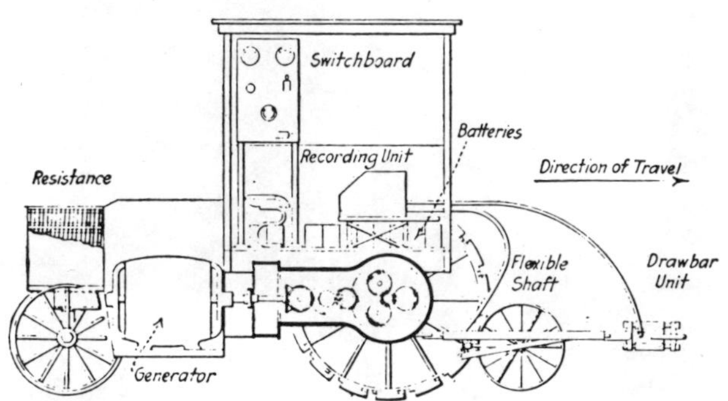

Figure 7
Testing car circa 1920 at the University of Nebraska

Figure 8
Testing car with instrument wagon, University of Nebraska, circa 1928

Figure 9
Load unit with instrument car, circa 1939

Figure 10
Load unit with instrument car, 1978

Figure 11
Testing course, University of Nebraska, 1979

Figure 12
Test car used in England in 1930's

Figure 13
Test car instrumentation used in England in 1930's

Table I

SAE/ASAE Code for Testing and Rating Tractors

Year	Brief Contents of Code
	(For full explanation obtain the pertinent copy of the code)
1917-1920	1. Belt pulley test at power tractor guaranteed to develop continuously over a two-hour period. 2. Drawbar test at power tractor guaranteed to develop continuously over a two-hour period. 3. Tractor to be rated on both the belt pulley and drawbar at 80% of the value determined above tests 1 and 2.
1920-1922	1. Code consists of rating system same as 1917-1920. 2. Brake horsepower tests as follows: a. Limbering-up run (up to 12 hours) b. Brake horsepower test at maximum load at rated engine speed. Carburetor set for maximum power with duration of test one hour. c. Brake horsepower test at 80% of maximum load recorded in (b). Test to be know as "rated load". Carburetor set to show fuel consumption at "rated load" and for a duration of two hours. d. Brake horsepower test at varying load. The object of this test is to show fuel consumption and governor control when load varies (all adjustments as in (c) above.) 3. Ratings a. Brake horsepower rating to be 80% of maximum belt horsepower (b) b. Drawbar horsepower rating to be (1) 60% of brake horsepower rating or (2) 80% of maximum drawbar horsepower developed in the maximum drawbar test under the rules of the Nebraska test (Nebraska Circular 13 April, 1921)
1923-1924	Small editorial changes made to bring wording similar to Nebraska as Bulletin 13, April 1921, plus the following additions: Sec. 1. Nature of tests: Section describes that three or more

Nebraska "Rules"

Year	Brief Contents of "Rules" for Official Tractor Tests
	(For full information obtain the "Rules Bulletins")
1919-1926	1. Brake horsepower test a. Limbering up runs. Duration of test up to 12 hours. b. Brake horsepower at rated load at rate engine speed. Object of test is to show whether tractor will carry continuously its rated load. Duration of test is for two hours, test later known as test (B). c. Brake horsepower test at varying load; namely, a sequence of rated, max., 0, ¼, ½, ¾ load. Object of test is to show fuel consumption and governor control when load varies. All adjustments are as of (B), later known as test (C). d. Brake horsepower test at maximum load. Object of test is to determine the greatest load tractor will carry on the belt for one hour at rated speed. Governor at full opening, carburetor adjusted to give maximum power, later known as test (D). e. Brake horsepower test at half load. Carburetor set to give most economical operation at this load. Duratin of test, one hour. later known as test (E). 2. a. Drawbar, horsepower test at rated load. Tests made on half-mile track, not level but has several short grades. Carburetor set for minimum fuel consumption with satisfactory operation. Gear to be that generally used for plowing. Duration of test to be ten hours. Later know as test (F). b. Drawbar test at maximum load. The object of this test is to determine the maximum drawbar pull which the tractor will develop on the track. Carburetor adjusted as in (1d) above, gear as in (2a) above, load applied until engine reaches rated engine speed. This test later becomes known as test (G). 3. Miscellaneous tests. Note: Neither the Nebraska law nor the Nebraska "Rules for Official Tests" call for a tractor rating; however, the reports included the SAE/ASAE tractor rating.

SAE/ASAE Code for Testing and Rating Tractors

Year	Brief Contents of Code (For full explanation obtain the pertinent copy of the code)
	tractors to be picked at random from factory stock and the average of all tests completed is to be used in determining the results. Sec. 4. Added to allow tests to be conducted at any place which meets approval of both the manufacturer and engineers who conduct the test. Drawbar rating at 60% of brake horsepower rating deleted.
1925	1. The tractor rating was revised as follows: a. The belt horsepower rating was revised 80 to 90% of the maximum load which the engine will maintain by the belt at the brake dynamometer for two hours at rated engine rpm; however, the load for the rated brake horsepower remained at 80%. b. The drawbar horsepower rating remained at 80% maximum drawbar horsepower. 2. A correction factor for temperature and altitude affect on horsepower was added. Standard temperature 70°F (530°abs T), barometer 28.6 inches of mercury. Formula to be as follows: $B\text{-}Hp_C = B\text{-}Hp_O \times \frac{(Ps)}{(Po)} \times \frac{(To)}{(Ts)}$* *If 60°F and 30 inches Hg is used, then belt pulley rating shall be 80% instead of 90%. Drawbar rating shall be 75% instead of 80%. 3. Test relettered to align and agree with those of Nebraska rules for official tractor tests. 4. Test "C" added for maximum drawbar horsepower. Tests to be run on a 1,000 foot course, preferably a "firm level track". (Note: Nebraska rules at that time call for one-half mile course not level but with short grades).
1926-1928	1. The belt horsepower rating of the tractor was revised to be 85% of the tractor maximum load maintained by the belt at the brake or dynamometer

Nebraska "Rules"

Year	Brief Contents of "Rules" for Official Tractor Tests (For full information obtain the "Rules Bulletins")
	Law Amended Senate File 172, 40th Session (in force March 25, 1921) Nebraska Tractor Test Law (Par. 5531-5543, The Compiled statues of Nebraska, 1922) pages 1751 and 1752
1926	The belt pulley horsepower and drawbar horsepower ratings were determined from the maximum belt horsepower and maximum drawbar horsepower values

SAE/ASAE Code for Testing and Rating Tractors

Year	Brief Contents of Code
	(For full explanation obtain the pertinent copy of the code)
	at rated engine speed. 2. The drawbar rating was revised to 75% of the maximum drawbar horsepower. 3. Other test conditions remain per those revised in 1925.
1929-1935	
1936-1948 (1937 SAE Handbook)	Cooperatively, members of SAE/ASAE from industry and the University of Nebraska Tractor Test Station, brought the SAE/ASAE Tractor Test Code up-to-date generally in line with the practice at the Nebraska Test Station. The tests were as follows: **Belt Tests** Test B, maximum belt horsepower Test C, varying load belt horsepower Test D, torque test **Drawbar Test** Test A, Limber up Test E, Maximum drawbar horsepower Test F, Drawbar fuel consumption Note: The drawbar test was changed to be conducted over 500 foot course instead of 1000 feet. The power correction factor specified at temperature 60°F, and barometer - 29.92 inch mercury. One carburetor setting at "most satisfactory" adjustment and maintained for all tests. Test C, "Varying Load Belt Horsepower Tests" changed to following schedule: Governor adjusted to rated speed at 85% maximum horsepower corrected; with six runs at following loads; 85, 75, 50, 25, 15 and 0 percent maximum horsepower corrected. Maximum torque readings at 115, 100, 85, 70, 60 and 50 of rated engine speed. A drawbar fuel consumption test was added -- this to be conducted at 75% maximum drawbar horsepower corrected. Instructions relative to selecting the test tractor were deleted.

Nebraska "Rules"

Year	Brief Contents of "Rules" for Official Tractor Tests
	(For full information obtain the "Rules Bulletins")
	and applying the 85% and 75% factors as specified in the SAE/ASAE codes. All the remainder of the above tests per the "Rules for Official Tractor Tests" established in 1919/1928 continue to apply.
1928	Varying Load Tests changed from 10 min to 20 min per each load; sequence: rated, 0, ½, max., ¼, ¾ load.
1928-1934	Prior to 1928, it was permissible to make carburetor settings for each individual load test. Beginning in 1928, one carburetor setting was adopted for all results and remained unchanged throughout the complete test. This was the "operating setting". In determining the drawbar and belt ratings, runs were made at a 100% maximum carburetor setting.[11]
1933	Nebraska Tractor Test Law revised, House Roll 526, 49th Session (1933).
1935	Tests changed to better designation on Individual Test Reports. Corrected power added to reports.
1936-1950	The rules were amended in 1936.[11] The following is a brief of the rules. For full details, see University of Nebraska Bulletin No. 313. Letter designation for tests appeared on reports but designation in rules and procedures did not appear until 1937 test session. 1. Limber-up run **Test A** - The first test is made on the drawbar and is known as the "limber-up" test. The principal object of this test is to take out the stiffness likely to be found in a new machine and give the manufacturer an opportunity to check the condition of the tractor and ascertain if all the parts are working normally. The tractor is operated at approximately one-third rated load for four hours, two-thirds rated load for four hours and full rated load for four hours. 2. Belt Tests **Test B,** 100 % Maximum Belt Test -All adjustments are made to secure maximum output from the engine at rated speed. The governor is set to hold the throttle valve to the extreme open position, the spark is set to give the best results, the manifold heat con-

SAE/ASAE Code for Testing and Rating Tractors		Nebraska "Rules"	
Year	**Brief Contents of Code**	**Year**	**Brief Contents of "Rules" for Official Tractor Tests**
	(For full explanation obtain the pertinent copy of the code)		(For full information obtain the "Rules Bulletins")
			trol, if present, is set at the most favorable position, and the 1936 carburetor is carefully adjusted to that point at which an additional amount of fuel gives no increase but less fuel decreases the power output at the rated speed of the engine. After uniform operating conditions are reched, this test, known as the 100 % maximum, is continued for two hours. The object of this test is to check or establish belt horsepower ratings. The results are used in arriving at the "calculated" rating appearing on the individual reports. The observed maximum horsepower as determined in this initial belt test did not appear in the published report of each test nor on the summary sheet of the annual tractor test bulletin from 1928 to and including 1934, except in those cases where the manufacturer elected to use to 100 % maximum power setting throughout the complete test. Test B has appeared on all of the individual test reports in 1935-1949 inclusive. **Test C,** Operating Maximum Belt Test -In case the manufacturer wishes to use a carburetor adjustment leaner than the 100% setting, a series of trial runs of 20 to 30 minutes at leaner settings are made and the manufacturer's representative is permitted to choose from these an "operating setting". This operating setting of the carburetor is the setting recommended by the manufacturer as being the most practial for general use and is used throughout the remainder of the test with the exception of the 100% maximum drawbar test in rated gear (to determine the drawbar rating) when the adjustment used is the 100% maximum power setting as obtained on the belt. The first run on this operating setting is a one-hour maximum power to be known as the "operating maximum belt test". the object of this test is to determine the maximum power developed and the fuel consumption at a carburetor setting that is practical for field operations. **Test D,** Rated Load Belt Test - The object of this test is to determine

SAE/ASAE Code for Testing and Rating Tractors		Nebraska "Rules"	
Year	**Brief Contents of Code**	**Year**	**Brief Contents of "Rules" for Official Tractor Tests**
	(For full explanation obtain the pertinent copy of the code)		(For full information obtain the "Rules Bulletins")
			whether the tractor will carry its rated load on the belt and to secure a record of fuel consumption and other operating data. The carburetor remains set at the operating setting. Rated load is determined in one of two ways. If the tractor is rated by the manufacturer, sufficient load is applied to develop rated horsepower at rated engine speed. In case the manufacturer either does not specify the rating of the tractor or, elects to accept the "calculated" rating, then rated load becomes 85% of the corrected maximum as calculated from the results of the 100% maximum belt test. **Test E,** Varying Load Belt Test -The carburetor setting, governor setting and ignition timing are the same as in the rated load test. The object of this test is to show fuel consumption and governor control of the engine speed when the load varies. It is composed of six 20-minute runs made in the following order: The first is rated load, which is the same load as the one-hour rated load test. This is followed by 20 minutes at no load. The minimum load is applied and the power developed is approximately one horsepower. The next load applied is one-half of the rated load torque and this is followed by a maximum horsepower test in which enough load is applied on the dynamometer to pull the engine far enough under rated speed to insure an extreme throttle opening. The last two runs are made at one-fourth and three-fourths rated load torque, respectively. 3. Drawbar Tests (The drawbar test course is for a distance of 500 feet) **Test F,** 100% Maximum Drawbar Test - The first drawbar test is a maximum horsepower test using the "100%" carburetor setting as found on the belt. The 100% maximum test is made in one gear only - that gear designated by the manufacturer as most suitable for plowing or ordinary farm work. This is commonly known as "rated gear". The results of this test are used to determine the "calculated" drawbar rating. The observed 100% maximum drawbar

SAE/ASAE Code for Testing and Rating Tractors		Nebraska "Rules"	
Year	Brief Contents of Code (For full explanation obtain the pertinent copy of the code)	Year	Brief Contents of "Rules" for Official Tractor Tests (For full information obtain the "Rules Bulletins")
			horsepower is corrected to standard conditions and multiplied by 0.75. **Test G,** Operating Maximum Drawbar Test - Operating maximum drawbar tests are made in the forward gears, using the operating carburetor setting selected and used on the operating belt tests. The object of this test is to determine the maximum horsepower the tractor will develop in each forward gear with engine running at rated speed under the prevailing temperature and barometric conditions. To determine the maximum horsepower in such gears, a series of runs is made in each gear, starting with the throttle partly closed so that the governor controls the engine speed and with sufficient load applied to produce a driven-wheel slippage of 5 to 6%. Other runs are made with increased load, but with the throttle opened wider to give rated engine speed which increases slippage. This balancing of throttle opening, engine speed, load, and slippage is carried to the point where either the engine is developing maximum power or the slippage is so great that the horsepower is appreciably reduced. Usually no results are used when the slippage exceeds 16%. **Test H,** Rated Load Drawbar Test -The next run made on the drawbar is the rated load test. The duration of the test is ten hours actual running time, as nearly continuous as possible, with constant load. A record is made of the time and reason for each stop, and also of any adjustments or repairs made on the tractor. The objects of this test are to determine whether the tractor can pull its rated load continuously and to secure a record of fuel consumption on drawbar work. The governor is adjusted to give rated speed of the engine with the tractor pulling rated load. The gear used is that recommended for plowing or ordinary farm work, which is known as rated gear. The carburetor remains at the operating setting. Rated load is determined in one of two ways. If the tractor is rated by the manufacturer, sufficient load is applied to develop rated

SAE/ASAE Code for Testing and Rating Tractors

Nebraska "Rules"

Year	Brief Contents of Code (For full explanation obtain the pertinent copy of the code)	Year	Brief Contents of "Rules" for Official Tractor Tests (For full information obtain the "Rules Bulletins")
			horsepower at rated engine speed in rated gear. In case the manufacturer either does not specify the rating of the tractor or elects to accept the "calculated" rating, then rated load becomes 75% of the corrected maximum as calculated from the results of the 100% maximum drawbar test in rated gear. **Test J** - is made in rated gear only, using the operating carburetor setting. The principal object of Test J is to show the effect of the removal of added weight on the performance of the tractor. In Test J, the wheel and the tire equipment used is the same as in Tests F, G, and H except that all added weight, either liquid, cast iron or any other form, is removed. Test J may be compared directly with the rated gear run in Test G. **Test K** - is made in rated gear only, using the operating carburetor setting. The principal object of Test K is to show the effect of using smaller tires and wheels on the performance of the tractor. In Test K the smallest wheels and tires furnished as optinal equipment by the manufacturer are used. This test can be compared with Test J. All added weight, either liquid, cast iron, or any other form, is removed. Note: The SAE/ASAE code as revised in 1936,made no provision for determining horsepower ratings. The Nebraska "Rules" continued with the same method as outlined in the former code.
1949-1955	In 1948, (published in October, 1948) the SAE/ASAE Tractor Code was revised. From the general instructions it would lead one to believe that the code established in 1937 was still to be used but recognition of the tractors being equipped with rubber tires. A section with description of instrumentation, a glossary of terms and specified formulas to be used were added.	1951-1958	A brief outline of the: (Back of each individual Test Report begins with No. 454.) 1 . Limber-up run **Test A** - The manufacturer's representative operates the tractor for a minimum of 12 hours using light to heavy drawbar loads in each gear. This serves as a preliminary period for limber-up, general observation and adjustments. Adjustments that are permissible include setting the valve tappet clearance and the gap of the breaker points or spark plugs, ad-
1956-1958	In 1955 (and approved 1956), the SAE/ASAE Tractor Test Codes (renamed "Agricultural Wheel Type Tractor Test Code"), was revised and updated again, bringing the code into agreement with the practices used at the Nebraska Trac-		

SAE/ASAE Code for Testing and Rating Tractors		**Nebraska "Rules"**	
Year	**Brief Contents of Code** (For full explanation obtain the pertinent copy of the code)	**Year**	**Brief Contents of "Rules" for Official Tractor Tests** (For full information obtain the "Rules Bulletins")

tor Test Laboratory. Outline of the code is as follows:

Preliminary Test
Test A - Break-in or limber-up.

Power Outlet Performance Test (Belt or power take-off)
Test B - 100% maximum power
Test C - Operating maximum power
Test D - Rated power
Test E - Varying power
Test L - Torque (at dynamometer)

Drawbar Performance Test
Test F - 100% maximum drawbar power
Test G - Operating maximum drawbar
Test H - Rated power
Test J - Power without ballast
Test K - Power without ballast and with smallest tires and lightest wheels regularly furnished for the tractor.

Major points of interest are:

1. For the first time, a "stock model" is defined as "equipped only with those items that are available to the customer". It is expected that the tractor being tested will be equiped with the most popular items.

1. Variable load test (varying power test) is again changed as follows: (Carburetor and governor set as per rated power)

   a. Rated power
   b. No load (minimum load on the dynamometer)
   c. One-half rated power (½ of rated torque)
   d. Governed maximum power
   e. One-fourth rated power (¼ of rated torque)
   f. Three-fourths rated power (¾ of rated torque)

3. Recognition of power take-offs.

4. Traction surface specified as "brushed concrete".

justing the clutch, and making other adjustments of a similar nature. No new parts or accessories can be installed without having mention made of it in the report.

No data are recorded during this preliminary run except the time that the engine is operated.

2. Belt Tests

Test B - The throttle valve is held wide open and the belt load on the dynamometer is adjusted so the engine is at the rated speed recommended by the manufacturer. Carburetor, ignition timing and manifold adjustments are all set for maximum engine power.

This test is designed to determine maximum belt horsepower of the tractor engine at rated speed and measure fuel consumption at the maximum power on the belt.

Test C - When tractors use carburetors the best fuel economy does not always occur when the engine develops maximum power at rated speed. Test C is intended to allow the manufacturer's representative to select a more economical fuel setting even though there is a slight loss of power. This more practical carburetor setting is used in all later tests except Test F. Throttle valve is held wide open and load adjusted to give rated rpm. Tests B and C are the same for Diesel tractors, which have an altogether different fuel system.'

Test D - The throttle control lever is set so the governor will maintain rated engine speed when rated load is applied. Rated load is 85% of 100% maximum, as obtained in Test B, corrected to standard conditions.

This rating is somewhat less than the maximum belt hp in order that the operator may have a certain amount of reserve.

Test E - Serves to show the range of engine speeds when the engine is controlled by the governor during the following varied loads of 20 minutes each: rated load, no load, ½ rated

SAE/ASAE Code for Testing and Rating Tractors		Nebraska "Rules"	
Year	**Brief Contents of Code**	**Year**	**Brief Contents of "Rules" for Official Tractor Tests**
	(For full explanation obtain the pertinent copy of the code)		(For full information obtain the "Rules Bulletins")

load, maximum load at wide open throttle valve, ¼ and ¾ rated load.

The average result of this test shows the average power and fuel consumption. Since the average tractor is subjected to varying loads, these data serve well in predicting fuel consumption and efficiency of a tractor in general use.

3. Drawbar Tests

Test F - Is a preliminary drawbar test, the results of which are used to determine the rated drawbar horsepower in Test H. The carburetor is set to develop maximum power as in Test B. The rated gear recommended by manufacturer as plow gear is used in this test. The drawbar load is adjusted to give rated engine speed.

Test G - Determines maximum drawbar horsepower in each gear when the carburetor is set for fuel economy as in Test C. The throttle valve is held wide open and the load is applied so that the engine runs at rated engine speed.

When operating in low gear it is not uncommon for the tractor to develop less drawbar horsepower than in rated gear because of excessive wheel slippage. When excessive wheel slippage occurs the load is reduced until slippage approches 16%*. When the load is reduced it is necessary to operate the tractor engine at part throttle and control engine speed by governor action.

Because of the limited space in this column it is not possible to show results for more than four gears. Except for the lowest and highest gears the drawbar results are usually similar.

*changed to 15% in 1958.

Test H - Is intended to test the ability of the tractor to run continuously for 10 hours at rated drawbar horsepower and to determine the fuel consumption during that time. Rated drawbar horsepower is 75% of 100% maximum drawbar horsepower (Test F), corrected to standard conditions.

SAE/ASAE Code for Testing and Rating Tractors		Nebraska "Rules"	
Year	**Brief Contents of Code** (For full explanation obtain the pertinent copy of the code)	**Year**	**Brief Contents of "Rules" for Official Tractor Tests** (For full information obtain the "Rules Bulletins")
			When operating at rated load the throttle control lever is set to maintain rated engine speed. This rating is less than maximum drawbar horsepower in order that the owner may have a certain amount of reserve. **Test J** - The tractor is operated in rated gear with all added weight removed. This test shows the effect of the removal of added weight on the performance of the tractor. When compared with Test G, removal of wheel weights generally increases wheel slippage and decreases drawbar horsepower. **Test K** - Speed-Pull Characteristic Started on Drawbar (1958) Note: Belt hp rating is 85% of 100% maximum. (Test B), corrected to standard conditions. Drawbar hp rating is 75% of 100% maximum. (Test F), corrected to standard conditions.
1959-1971	Because of changes in style of transmission and power take-offs, changes in the SAE/ASAE Test Code were required. Outline of this is as follows: 1. Limber-up and preparation for performance test. 2. Mechanical power outlet a. Maximum power-fuel consumption b. Varying power-fuel consumption c. Power at SAE/ASAE standard power take-off speeds 3. Drawbar performance a. Maximum drawbar power with ballast b. Varying drawbar power-fuel consumption with ballast c. Drawbar pull versus travel speed with ballast d. Maximum drawbar power without ballast Major points of interest are: 1. The torque test at the PTO shaft dynamometer was deleted.	1959-1970	Brief outline of test is as follows: Same PTO tests basically but letter designation dropped. No correction factors. 1. Limber-up run The tractor is operated in each gear with light to heavy loads during the limber-up period of twelve hours. Any parts added or repaced during the limber-up run and subsequent runs are mentioned in the test report. 2. Power outlet performance test Power outlet performance runs are made by connecting either the belt pulley or the power take-off to a dynamometer. During a preliminary power outlet run the manufacturer's representative may make final adjustment for the fuel, ignition and governor control settings. The manually operated governor control mechanism is set to provide the high-idle speed specified by the manufacturer for maximum power. No adjustments or settings are changed during all subsequent runs. Power consuming accessories may be disconnected only when it is con-

SAE/ASAE Code for Testing and Rating Tractors

Year	Brief Contents of Code
	(For full explanation obtain the pertinent copy of the code)

2. A "lugging run" (torque test as described by drawbar pull was added in the test. Drawbar pull versus travel speed with ballast)
3. The correction factor (F) for power was deleted.
4. The stability when conducting drawbar tests was added.
5. The schedule of percentages of the "maximum power" at which the varying power-fuel consumption run where changed to again "correspond to the present practice".
6. A test of varying drawbar power-fuel consumption was added.
7. Recognition of characteristics of torque converter transmissions.
8. The length for drawbar tests was changed from 500 feet to 300 feet (minimum).
9. Minor editorial changes were made in 1961, 1964 and 1968.

Nebraska "Rules"

Year	Brief Contents of "Rules" for Official Tractor Tests
	(For full information obtain the "Rules Bulletins")

venient and recommended for the operator to do so in regular practice.

During the power outlet runs an ambient air temperature of 75°F is maintained.

a. Maximum Power and Fuel Consumption
Maximum power is obtained with either wide open throttle or the specified setting of the diesel pump and the rated engine speed specified by the manufacturer. Time of run is two hours. (Formerly Test C).

b. Standard Power Take-off Speed
Whenever the power take-off speed during the maximum power run differs from the speeds set forth in the ASAE and SAE standards an additional run is made at either 540 or 1000 rpm of the power take-off. Time of run is one hour.

c. Varying Power and Fuel Consumption
The engine torque is varied at 20-minute intervals in the following pattern: 85% of the engine torque at maximum power (formerly Test D), minimum dynamometer torque, ½ of the 85% torque, torque at the specified engine speed, ¼ of the 85% torque and ¾ of the 85% torque. Total time of run is two hours. (Formerly Test E).

The "Average" results of this run indicate the "average" performance of a tractor under the varying loads encountered in field operations.

3. Drawbar performance Test (65% Lug Height) ASAE Paper 58-604 or Transactions Article
Maximum drawbar power is determined at the manufacturer's specified engine speed in each gear of the tractor within the following limits: (1) slip of the drivers may not exceed 15% for pneumatic tires on the concrete test course or 7% for steel cleats on the well-packed earthen test course, (2) ground speeds may not exceed 15 miles per hour, (3) safe stability limits of the tractor may not be exceeded, (4) static tire loads and inflation

SAE/ASAE Code for Testing and Rating Tractors

Year	Brief Contents of Code
	(For full explanation obtain the pertinent copy of the code)

Nebraska "Rules"

Year	Brief Contents of "Rules" for Official Tractor Tests
	(For full information obtain the "Rules Bulletins")

pressures must conform to published standards, (5) other operating limits.

Drawbar load is increased until the manufacturer's specified engine speed is obtained with either wide open throttle or the specified setting of the diesel pump. Travel speed, drawbar pull and other data are recorded over 500-foot straight level portions of the test course.

The maximum drawbar horsepower and the corresponding speed and drawbar pull for up to 12 gears or travel speeds are shown in the individual test reports.

a. Maximum Power
Maximum drawbar power is shown for the normal travel speed selected by the manufacturer. (Formerly Test G).

b. Maximum Pull
Maximum pull is applied to cause decreasing travel speeds. The pull is limited by either engine power, slippage, stability or some other operating limit.

c. Varying Power Tests (Fuel measured in these tests)
The varying power runs are made to show the effect of speed control devices (engine governor, automatic transmissions, etc.) on horsepower, speed and fuel consumption. These runs are made around the entire test course which has two 180 degree turns.

(1) Maximum Available Power
The tractor is operated with a drawbar pull such that the manufacturer's selected travel speed is maintained over the straight 500-foot sections of the test course. Power output during this run may differ somewhat from that obtained in the maximum power run because of load changes around the two 180 degree turns. Time of run is two hours.

(2) 75% of Pull at Maximum Power - The tractor is operated at 75% of the drawbar pull obtained

SAE/ASAE Code for Testing and Rating Tractors

Year	Brief Contents of Code
	(For full explanation obtain the pertinent copy of the code)
1972 to date	The SAE/ASAE Agricultural Tractor Test Code was revised by addition of test for noise at both "Operator's Station" and Drive by". Outline of code is as follows: 1. Limber-up and preparation of tractor for performance runs. 2. Mechanical power outlet performance a. Maximum power-fuel consumption b. Varying power-fuel consumption c. Power at SAE and ASAE Standard power take-off speed 3. Drawbar performance a. Maximum drawbar power with ballast b. Varying drawbar power-fuel consumption with ballast, including sound level at Operator's Station. c. Drawbar pull versus travel speed with ballast. d. Maximum drawbar power without ballast. e. Exterior sound level. Note: Major change was the addition of Noise Test at both the "Operator's Station" and "Driven By". The procedure outlined in the ISO Draft Proposal in TC 43/SC1, later known as ______________.

Nebraska "Rules"

Year	Brief Contents of "Rules" for Official Tractor Tests
	(For full information obtain the "Rules Bulletins")
	during the maximum power run. Time of run is 10 hours (formerly Test H). (3) 50% of Pull at Maximum Power - The tractor is operated at 50% of the drawbar pull obtained during the maximum power run. Time of run is 2 hours. Note: 1. The SAE/ASAE ratings for belt pulley and drawbar were deleted in 1959. 2. The use of the SAE power correction factor was deleted in 1959. 3. The schedule for percentages of "maximum power" at which the varying power-fuel consumption was changed. 4. The unballasted drawbar test was deleted. 5. Test to determine tractor's lugging ability added. 6. Test K revised, 1958. 7. Limitation on tractor tire lug height at completion of test (65% original height) added (1958).
1972 to date	Test schedule above (1950-1978) remained as is with the addition of the following: 1. A run at 50% maximum drawbar pull at reduced engine speed was added. In this test the tractor is shifted to a high gear (or speed rate and the engine speed is reduced until the same travel speed and pull obtained as recorded in the above test at 50% of pull at maximum power. 2. A test for the sound level generated a. At the tractor's operator station with the tractor in motion and under load. b . At a point 7.5 meters from the tractor with the tractor in motion and no load.

Table II
Chain of Events Leading to the SAE Agricultural Tractor Test Code (SAE J708c/ASAE S209.4) of Today

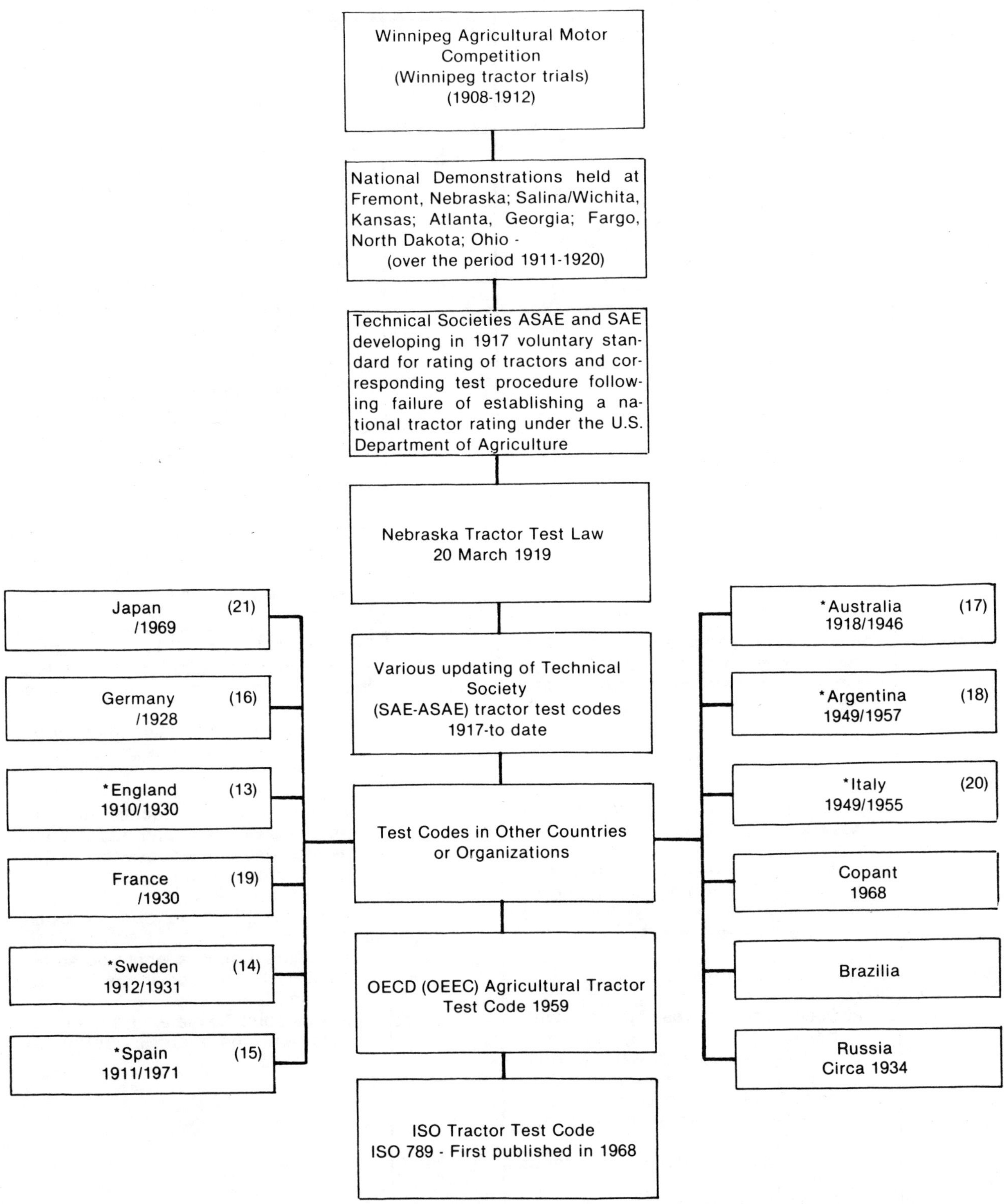

*Where two dates are shown, the first date was when the drawbar test portion of the test code was conducted under field plowing operation - measuring the resulting drawbar pull. The second date was when drawbar tests were conducted on a test track (course).

Table III
Nebraska Tractor Testing Law

Initial Enactment -	March 20, 1919. First effective - July 15, 1919. Introduced by Representative W. F. Crozier of Polk County, Nebraska. (House Roll No. 85)
Minor Amendment -	March 25, 1921. Tractor Test Fee of $250 per test. Introduced by Senator Chas. J. Warner, a farmer near Waverly, Nebraska. (Senate File No. 172)
Minor Amendment -	May 10, 1933. Tractor Test Fee of $500 per test. Introduced by Representative W. F. Crozier of Polk County, Nebraska. (House Roll No. 526)
Amendment -	April 9, 1949. Authorizing the Board of Regents of the University of Nebraska to fix the fees for testing tractors. Introduced by Thomas M. Davies of Lancaster Country, Nebraska. (Legislative Bill 310)
Amendment -	April 19, 1961. Exclusion of Tractor Less Than 10 hp (Legislative Bill 350)
Amendment -	July 12, 1963. General revision of the laws administered by the State Railway Commission. (LB 82). First exemptions written into the law - Tractors of less than 10 hp and truck tractors. Introduced by Schmit, Chief Eng. for the Railway Commission. (Legislative Bill 82)
Amendment -	July 19, 1965. Exemption for Heavy Construction and Earth moving Tractors. Introduced by Albert A. Kjar, 39th District; W. H. Hasebroock, 16th District (Legislative Bill 552)
Amendment -	June 8, 1967. Transfer of jurisdiction to Department of Agriculture and change exemption from 10 hp to 20 hp and of one cylinder. Introduced by Rick Budd, 2nd District; Albert Kjar, 29th District; C. W. Holmquist, 16th District; Jerone Warner, 25th District. (Legislative Bill 404)
Amendment -	May 18, 1971. Placing emphasis on tractor **engine**. Testing for all regular power outlets and health and safety tests. Other minor amendments. (Legislative Bill 692)

REFERENCES

1. Development of the Agricultural Tractor in the United States compiled by R. B. Gray -Parts 1 and 2.

   A. Page 3
   B. Page 47
   C. Page 49
   D. Page 129
   E. Page 134
   F. Page 1, part 2

2. Brief of draft of rules for Agricultural Motor Competition of Canadian Industrial Exhibition - Winnipeg, Canada, 1910.

3. "Nebraska Tractor Tests" 1917 - L. W. Chase, ASAE, Vol. XI, 1917, pages 132-158.

4. a. Minutes of Council Meeting of Society of Automobile Engineers, December 14, 1916; and, Society of Tractor Engineers, December 9, 1916. (SAE Bulletin, December, 1919, page 332).

   b. Meeting Report of Society of Automotive Engineers, Vol. I, August 1917, No. 2, Pages 162-163.

5. Standard Ratings for Tractors - Vol. III, No. 3, September, 1918.

6. a. Implement and Tractor Journal, September, 1919, W. F. Crozier - "Why a Nebraska Tractor Test Bill?"
   b. Nebraska Tractor Tests - Farm Implement News, September 25, 1952; October 10, 1952; October 25, 1952; November 10, 1952; November 25, 1952.
   c. "Tractor Makers Seek Official Rating" -Farm Implement News, April 17, 1919, plus newspaper and magazine clippings.

7. Legislation of Nebraska, 37th Session, House Roll No. 85; committee ammendments to House Roll No. 85; Chapter 218 House Roll 85 as signed by the Governor of Nebraska, March 20, 1919.

8. a. University of Nebraska Agricultural Experiment Station Circular No. 10, December, 1919, and Circular No. 13, April 1921.
   b. SAE Transactions, 1920, Part I, pages 635-644.

9. SAE Engine Test Code 1917, Vol. III, December, 1918, No. 6.

10. University of Nebraska Agricultural Experiment Station Bulletin 233, 1929.

11. University of Nebraska Agricultural Experiment Station Bulletin 397, January, 1950.

12. SAE and ASAE papers

    a. ASAE paper - E. E. Brackett, 1931 -The Nebraska Tractor Tests
    b. SAE paper - L. F. Larsen (September 10-13, 1956), Nebraska Tractor Test Charges.
    c. ASAE paper - L. W. Hurlbut, et al, ASAE No. 1005, December 17, 1959. Nebraska Tractor Test in Relation to SAE/ASAE Tractor Test Code.
    d. SAE paper 660584, W. H. Worthington, 50 years of Agricultural tractor development.
    e. SAE paper 730762, Manby, et al, "Development and Operation of OECD Tractor Test Code".
    f. SAE paper 730763, W. E. Splinter, et al, "Nebraska Tractor Test - Programs and Philosophy".
    g. ASAE paper - 1976, L. F. Larsen and Dr. L. Leviticus, "History of Nebraska Tractor Test".

13. "The Measurement of Tractor Performance", T. C. D. Manby, Institute of Mechanical Engineers, July 26, 1961.

14. Profningar of Maskiner och Redskap for Landtbrukets Behof - Middlande, Nos. 35, 58, and 60.

15. Spain Estacion de Ensayo de Maquinas, Instituto Agricola De Alfonso - D. Jose De Arce - 1911.

16. Land Technische Forschung Volume 1951-1954, Issue 21 - 1953, Munich Page 41.

17. Australia - Tractor Testing on Australia - G. H. Vasey and W. F. Baillie, current program of the Werribee Tractor Testing Station, W. F. Baillie, 1971 National Agricultural Machinery Conference.

18. Evolucion de Los Ensayo de Tractores en

La Argentina - lettre Jose Maria Casares, Secretaria de Estado de Agricultura y Grandaderia to Ing. Casiano L. C. Asas John Deere Agrentina, 25 de Juneo de 1979.

19. Letter, S. Picker, CNEEMA (Antony France) to F. C. Walters (John Deere Waterloo, PEC)

20. Letter C. Candini (Fiat Trattori - San Matteo, Modeno Italy Italy) to F. C. Walters (John Deere Waterloo, PEC)

21. Letter M. Karato (Yanmar Diesel Engines, Osaka, Japan to F. C. Walters (John Deere Waterloo, PEC)

Acknowledgements: The writers wish to express appreciation for the assistance they have received from Mr. Ralph Palmer of ASAE; Mr. G. LaLiberte, University of Winnipeg, Winnipeg, Canada; the library personnel from the SAE, ASAE, University of Wisconsin (Madison), University of Nebraska, C. Y. Thompson, Deere & Company, and John Deere Product Engineering Library in gathering the data. They further wish to acknowledge the help and information gained from interviews with Mr. W. H. Worthington (Wayne Engineering of Cedar Falls, Iowa); the former test engineers in charge of the Nebraska Test Laboratory, Mr. Carlton Zink and Mr. Lester Larsen; the present test engineers in charge, Dr. L. Leviticus and Dr. K. von Bargen; J. Matthews (NIAE), O. Nordstrum (Swedish Agricultural Equipment Test Station), A. Ynat (John Deere Iberica), and Andrian A. Andreas (Chamberlain - John Deere); and W. Brown of Australia Test Station; and other people who answered the writers' many questions. We wish also to express our appreciation to the personnel of John Deere Product Engineering Center who participated in the preparation of this paper.

800939

The Patent System and Agricultural Equipment

John A. Pekar
U. S. Patent and Trademark Office

THE PURPOSE OF THIS PAPER is to present to the technically trained person a historical perspective of the patent system with its general relationship to agriculture and farm equipment.

Today the patent system is accepted both by the patent practitioner and the inventor as a complicated composite of legal and technical procedures which require specialized training and experience to interpret. Its function and value to the individual who attempts to use it are sometimes obscured by its very complexity. By examining it from a cultural basis rather than from a legal, technological, or economic basis, its function and value may be clarified.

Emphasis is placed on its earlier history to better show its effects on the individual. Historically prominent persons are introduced to show the importance attached to the patent system by the Government and by individuals involved in earlier patent practice.

EVOLUTION OF THE PATENT SYSTEM

In Mark Twain's novel, "A Connecticut Yankee in King Arthur's Court," the inventive genius who becomes the actual ruler of Camelot casually remarks that "the very first official thing I did in my administration -- and it was on the very first day of it, too -- was to start a patent office; for I know that a country without a patent office and good patent laws was just a crab, and couldn't travel any way but side-ways or backwards." This allegorical reference to our patent system was made at a time when it evolved essentially into the system we have today. If a patent system were introduced into sixth century Britain, it is not clear whether it would have raised Camelot to the high technological level described in the novel. In order for a patent system to work in any country, a cultural bond must exist between the inventor and society. The inventor must be capable of studying and copying various devices in an art field, and through the process of emulation, be capable of improving them. The term emulation does not mean mere copying itself, but also the desire to excel. It was by emulation that the journeyman craftsman became a master himself. Society must recognize this innovative effort and encourage it. This recognition and encouragement did not exist in sixth century Britain. However, as pointed out by historian Brooke Hindle in his lecture entitled "Emulation and Invention" they did exist in the American colonies during the eighteenth century.

Our patent system did not start with a body of laws and procedures universally recognized by all as inherently good, nor was it instantly created as in Twain's tale. It was a cultural phenomenon, rather than a legal creation, which gradually evolved from the forms of recognition and encouragement of invention which preceded it. It was necessarily modified from time to time to better perform its function.

Any attempt to show the relationship between our patent system and farm equipment must necessarily consider the colonial period to examine what technology existed, how it was recognized and encouraged, and how it related to the farmer. By 1776 shipbuilding was the most successful colonial industry, and about a third of Britain's ships were built in America. Other important

ABSTRACT

A historical survey of the patent system is presented with special emphasis on how and why it was established, how it related to the farmer and farm equipment, and how it changed to keep pace with technological growth in the United States.

SAE/SP-80/470/$02.50

industries of the period were iron making and textile manufacturing. America had its craftsmen and mechanics who were adept at use of machine tools as evidenced by colonial clockmaking and gunsmithing. Societies were established in Boston, New York, Philadelphia, Baltimore and Charleston to encourage the improvement of manufacturing techniques and invention. This was accomplished by awarding prizes for significant inventions, or in some cases by the awarding of grants to finance the development of an invention. However, the chief benefit derived from the societies was educational. The American societies received models of inventions made in Britain, which could be examined by American craftsmen and mechanics. Thus, there was a transfer of British technology to the colonies, which could be copied and improved upon. Technology was transferred in other ways; immigration from Britain and the rest of Europe brought it here from different cultural backgrounds. Some Americans traveled to Europe to study a particular technological area, and frequently brought back experts in that technology to work in the colonies. Although the earliest transfers of technology were primarily from Britain, America was exposed to the sum total of European technology. For example, at the time of the Revolution, British, German and Flemish iron making processes were simultaneously used. The societies also maintained displays of models, machines, and machine tools for encouraging youths to become mechanics, and frequently taught drawing courses to stimulate their interest. The colonial farmer was exposed to technology, and knew mechanics and craftsmen. Sawmills and grist mills were common in the colonial period, and the water wheels and gear trains which powered them were probably a source of wonderment. Every farmer owned, or at least knew of, the Pennsylvania or Kentucky rifle which was superior to the military smooth bore muskets of the period. The farmer knew the country blacksmith who made and repaired the existing farm implements. Frequently, the farmer had to become his own mechanic and blacksmith, so that he became an all around handyman, a trait which is retained today. The handyman farmer along with the blacksmith and mechanic became the folk inventor which predominated our early history.

The first patent on machinery was issued in the colonies in 1646. However, many colonies never issued patents, and they did not play any significant part during the colonial period. The patents that were granted during this period were not issued by a governmental agency, but rather by acts of a colonial legislature.

In the period following the Revolutionary War, when the United States was governed by the Articles of Confederation, several states issued their own patents, and dispensed prizes and grants for useful inventions. During the Constitutional Convention the need to encourage and reward invention was recognized and debated. However, strange as it may seem, Benjamin Franklin, the most knowledgeable American in scientific and technological matters, did not appear to speak on this issue. Among the proposals set forth was the awarding of prizes and grants by the government. The main opposition to this proposal was that the money for the prizes and awards would have to be raised by taxes. As suggested by Brooke Hindle, it appears that the patent system with the reward of the market place to encourage and recognize invention was adopted simply because it was the cheapest method. Our patent system was created by the following clause inserted into Article 1, Section 8 of the Constitution: "Congress shall have power ... to promote the progress of science and the useful arts by securing for limited times to authors and inventors the exclusive right to their respective writings and inventions." Therefore, under our patent system a patent is a legal document securing for an inventor the exclusive right to make, use and sell his invention for a term of years. In return for patent rights, an inventor had the incentive to disclose his invention and thus further promote technological progress, rather than keeping his invention secret.

The idea of awarding prizes and grants did not die out. In 1849, in his Annual Report, Commissioner of Patents, Thomas Ewbank proposed that a fund be set up to award cash rewards every four years to the most important addition to science and the useful arts. He also proposed that medals be awarded to inventors. However, neither of his proposals was accepted. Henry Ellsworth, Commissioner of Patents, 1836-1845, helped Samuel Morse to obtain a $30,000 grant from Congress for testing the possibilities of telegraph.

FOUNDING OF PATENT SYSTEM

The United States patent system was founded on the Patent Act of April 10, 1790. Although President Washington in his first message to Congress urged that foreign inventions also be given patent protection, his advice was not followed, and patents could only be granted to citizens. The Act established a Patent Board of three persons: the Secretary of State, Thomas Jefferson; the Secretary of War, Henry Knox; and the Attorney General, Edmund Randolph; to issue patents for sufficiently useful and important inventions and discoveries. The last clause of the Act was held to give the Board power to refuse to issue patents for lack of novelty and insufficient utility or importance. The Board exercised its power of rejection so vigorously that in its three years of existence, it only issued forty seven patents. There was no appeal from the decisions of the Board which led to the charge against it, that by the very nature of its members, it was hostile to the inventor who represented the industrial class.

The Patent Act of 1793 abolished the Patent Board and eliminated the requirement of the previous Act that an invention be "sufficiently useful and important" which eliminated the rejection practice utilized by the Board. The Act gave the United States a registration system in which a patent was examined as to novelty and priority only when it was involved in litigation in the courts. The registration system lasted until 1836 and had many deficiencies, and at best, was run in an unbusiness-like manner.

The Patent Act of July 4, 1836 completely changed the character of the preceding system, and although statutes, rules, and procedures have since been changed and modified, it is essentially the same system we have today. The Act provided for an examination of every application so as to preclude granting duplicate patents to different parties for the same invention which happened frequently in the past. It set forth novelty and usefulness as criteria for patentability. By requiring the Commissioner and the Chief Clerk, who was in effect the Deputy Commissioner, to post bonds guaranteeing the faithful performance of their duties, it brought to the Patent Office a professionalism which in some cases was lacking under the preceding registration system. The Act provided procedures for determining priority between interfering applications, that is applications by different parties claiming the same invention and filed at substantially the same time. It provided a procedure for hearing appeals from adverse decisions of examiners. The term of the patent was fourteen years, and a procedure was provided for granting a seven year extension if the patent was not remunerative within its original term.

The Patent Office served agriculture in ways other than by issuing patents. In 1839, the Commissioner of Patents was given the additional duty of collecting and publishing statistics and other information on agriculture. Until the Department of Agriculture was established in 1862, the Commissioner was the de facto Secretary of Agriculture. The Patent Office published the first twenty Yearbooks of Agriculture, and part of the Office funds were used to distribute free seeds to farmers.

THE PATENT SYSTEM AND THE INTRODUCTION OF MECHANIZED AGRICULTURAL EQUIPMENT

It is most unfortunate that the examination system instituted in 1836 was not introduced earlier since significant changes occurred in American farming and technology during the early 1830's. The westward movement of the farmer to midwest and plains states was a reality. Huge tracts of prarie land could be acquired at modest prices. The ancient curse of the farmer, namely his ability to sow more than he could reap, was compounded. At this time grain throughout the world was harvested almost exclusively with the cradle or the scythe. It was estimated that a strong man with a cradle could cut three acres a day, and if a scythe were used somewhat less. Each tool also required at least one other person to rake and bind the stalks. The need for a mechanized harvesting device was recognized before this time, both in this country and in England. There were about forty five patents issued in both countries for harvesting devices. However, none of them achieved commercial success. On December 21, 1833, a patent was issued to Obed Hussey for a reaping and mowing machine, and on June 21, 1834, a patent was issued to Cyrus McCormick for a reaper.

I have chosen these two inventions for detailed discussion because they were the most controversial inventions in the farm equipment field, and they occurred in a time when the whole fabric of American life was changing. Through an examination of the tribulations of Hussey and McCormick we can appreciate what changes occurred in the country, farm equipment manufacture, the patent system, and economics.

Both Hussey and McCormick were Easterners which is not surprising since at this time Virginia was still the biggest wheat producing state. Both men fit the classic mold of the American folk inventor. From a letter written by his wife, we get the following picture of Obed Hussey. He was born in New England, and lived as a youth in the whaling town of Nantucket, Massacusetts. He made at least one voyage in a whaling ship, and possibly two. He was a competent draftsman and mechanic. He lived in Baltimore for twenty three years where he operated a corn grinding mill for a time, and besides his reaper, he invented a steam plow, a machine for grinding hooks and eyes, and a rubber block for tackle especially for use on ships, and he experimented in making artificial ice. He wrote peotry and prose, and was a literary acquaintance of Edgar Allen Poe.

McCormick was born and raised in Rockbridge County in western Virginia. His father, Robert McCormick was also an inventor, and among his inventions were a hemp brake, a bellows, a clover huller and a threshing machine. He was interested in reaping machines, but after expending much time and effort he was unsuccessful in his attempts to build a machine which worked properly. He was a prolific reader and a student of astronomy. Cyrus McCormick attended the local school, and worked both on his father's farm and in his blacksmith shop. His interests in the reaper were the results of his father's interests, and his personal experience as a farm laborer. At the age of fifteen, he made his own cradle on a reduced scale which was easier to swing, thereby reducing his own drudgery when harvesting wheat. At eighteen he studied surveying. However, his overriding interests centered on the development of the reaper. He completed his reaper in 1831, and on July 25th he held a public trial of the reaper near Steele's Tavern, Virginia. As a result of this trial, he received everlasting fame as the inventor of the first machine to cut grain successfully. Today he is universally recognized as the inventor of the reaper.

Neither Hussey's nor McCormick's machines proved to be an instant commercial success. Each device had to be subjected to numerous field trials in order to convince farmers that they would work properly. Each machine broke down at certain trials and worked well at others.

The field trials presented enormous difficulties in that the wheat harvesting period in any given area lasted from four days to a week, and due to the primitive means of transportation at the time, the trial period in every year was extremely short. Each machine received numerous testimonials. However, the vast majority of testimonial writers did not buy reapers. Both Hussey and McCormick were in a similar financial condition, and neither had nor could raise the necessary capital to properly exploit his patent. Although Hussey sold a few of his machines, he was far from a financial success, and McCormick did not sell one of his machines until 1840. Probable reasons for the failure to sell reapers were the economic havoc created by the Panic of 1837, and the low price of wheat coupled with a surplus of farm labor.

During the early 1840's the supremacy in wheat production had shifted westwardly, and after a slow start the use of reapers for harvesting increased. McCormick realized that the transportation of reapers to the West presented a complex problem since delivery in time for the harvesting season for which they were ordered could not be guaranteed. He licensed others to build his machines in various states in order to solve the transportation problem. However, this did not prove satisfactory since some reapers were built of inferior materials and delivery schedules were not met. Deciding that he must build his own machines, he in partnership with another, built a factory in Chicago in 1847 where he could rely on the Great Lakes transportation system both to supply him with materials from the East and to ship back reapers ordered from there. McCormick's reaper sales then increased dramatically so that he was able to buy his partner out within two years and run his business by himself. Hussey, on the other hand, relied on others to build his machines and never did achieve financial success. His leading proponent in the subsequent controversy regarding the actual inventorship of the first successful reaper, John F. Steward, stated, "No one will claim that Mr. Hussey was a good business man."

There is no evidence in the Patent Office records that Hussey applied for an extension of his 1833 patent, and his patent expired at the end of the original fourteen year term. McCormick applied for an extension of his patent in 1847. One of the provisions of the extension procedure was that patents issued before the 1836 Act must be subjected to an examination before the extension could be granted. Examiner Charles G. Page found that McCormick's cutting apparatus was found in the Hussey patent of 1833, and that his revolving rack was similar to that of James Ten Eyck patented in 1824. In order to secure an extension of his patent, McCormick had to prove priority of his invention over that of Hussey. A Patent Extension Board consisting of James Buchanan, Secretary of State, who later became President of the United States; Edmund Burke, Commissioner of Patents; and R.H. Gillet, Solicitor of the Treasury; permitted McCormick to take testimony regarding priority of invention. Since Hussey's patent was involved, he was given the right to cross examine witnesses. McCormick's testimony was not deemed persuasive by the Board. However, no reasons were given for their decision. The Board merely stated that, "having examined the evidence adduced in the case, decide said patent ought not to be extended." Both Hussey and McCormick petitioned Congress to extend their patents. However, after much controversy, the matter lapsed and neither patent was extended. The controversy regarding who actually invented the first reaper began. The market for reapers seemed almost limitless, and a dominant patent ensured riches to the inventor. However, there was also a fear that a gigantic monopoly would be created which would enslave the farmer. Congress was bombarded by pleas from the supporters of each man who praised one and deprecated the other, and by pleas from those who believed that they were in collusion, and would split the market between them if either received an extension. Since neither patent was extended, it is impossible to say whether or not the feared monopoly would have resulted.

Before McCormick applied for an extension, Hussey did not believe that his machine was similar to McCormick's. McCormick felt that he could beat Hussey in the market place. The controversy regarding priority of invention did not abate. McCormick desired recognition as the first inventor. However, this should not be interpreted as a character defect because as suggested by Brooke Hindle, it would amount to an extension of the principle wherein society must recognize the successful inventor. Society, realizing that there was much work to be done in the country, demanded recognition of the successful inventor, and held him out as someone to be emulated. An example of this is the engraving entitled "A Notable Group of American Inventors", by John Sartain, wherein McCormick is pictured along with others such as Samuel Colt, Peter Cooper, Joseph Henry, Charles Goodyear and Samuel Morse. McCormick is the sole representative of inventors of farm equipment. A model of the reaper appears prominently in the left bottom corner. Inasmuch as inventors of farm equipment are concerned, John Deere, who also fit the classic mold of the American folk inventor, could also have been included in the engraving. In advertisements for McCormick's reapers the first public trial near Steele's Tavern, Virginia was frequently illustrated and described, which rankled his competitors. Any determination of who was actually the first inventor would require a basic examination of the original papers of McCormick's priority testimony in his application for extension of his original patent, since the supporters of each man had their own axes to grind and their bias is easily discerned.

McCormick vigorously defended his subsequent patents and brought infringement suits. Numerous competitors entered the reaper market, and the basic device quickly improved. In the majority of infringement suits only claims drawn to specific improvements were held valid.

One infringement suit brought by McCormick against his most successful early competitor, John H. Manny, is of historical significance, not because of the result since McCormick's patents of 1847 and 1853 were held valid, but not infringed, but because of the principals who were involved in the presentation of the case. In 1855, McCormick instituted an infringement suit against John H. Manny and his associates in the Federal Court at Chicago. Manny's patent attorney, Peter Watson, who procured most of his patents, was given the responsibility for the defense, and he first secured the services of George Harding to argue the case. Since it appeared that the case would be heard in Chicago, an Illinois attorney, Abraham Lincoln, was hired as a local associate. Lincoln mistakenly believed he was hired to argue the case so he prepared his presentation. By agreement of both sides, the trial was moved to Cincinnati, Ohio where Lincoln went to present his defense of Manny's patent. He was side-tracked and ignored by his fellow counsel, and did not participate in the arguments. Lincoln received a fee of $1,000 for his services, and it is suggested that this fee financed his appearance during the Lincoln, Douglas debates which led to his nomination for President, and his election to that post. Edwin M. Stanton became the leading counsel for the Manny interests, and was later named Secretary of War by President Lincoln. George Harding was offered the post of Commissioner of Patents by Lincoln which he refused. Peter Watson became the president of the Erie Railroad. Appearing for McCormick were William H. Seward who was appointed Secretary of State by Lincoln, and Reverdy Johnson who became a Senator from Maryland during the Lincoln administration.

EARLY USE OF PATENT LICENSES

Patents were of use not only to an inventor or his assignees, as a consideration of the business practices of Jerome I. Case will show. In 1842 he traveled west with six primitive threshing machines called "groundhogs," and he sold five machines along the way to provide him with a grubstake. He settled in Racine, Wisconsin, where he designed and manufactured threshing machines. He was not actually an inventor, but rather a developer. His first machines were improvements on the "groundhog." He licensed patents of others and incorporated their features into his machines. He continued this practice when he expanded into other lines of farm equipment.

CHANGES IN THE INDUSTRY AND THE PATENT SYSTEM

The production of harvesters along with other types of machinery increased. In 1871, 160 patents were issued for reapers, 160 for plows and their attachments, 72 for threshers and 23 for corn pickers and huskers. Although the number of patents for farm machinery increased as competition increased, no patents ever caused the controvery occasioned by the first Hussey or McCormick patents. As pointed out by Paul C. Johnson in "Farm Inventions in the Making of America," the inventive environment in this field underwent a significant change. The importance of the innovative mechanic, farmer and country blacksmith as invantors and developers of farm equipment diminished because of the huge capital requirements for setting up manufacturing facilities and product development. A new type of inventor, the engineer, gradually arrived on the scene. This led the full line companies into taking over most of the inventive and development functions for improvements and new products. The mechanic, farmer and blacksmith still made inventions. However, they either had to sell their inventions or establish small businesses which produced specialized equipment. Such small line business proliferated, and some survived the fierce competition in the field and succeeded. Although there has been controversy regarding the full line companies buying up inventions or manufacturing parallel lines, the intense competition has produced the gigantic industry we have today. The patent helped this competition by securing for inventors their rights to their inventions.

As in the past, the Patent and Trademark Office is changing to meet the needs occasioned by changing technology and economics. The Office is now providing International Search Reports in applications filed under the recently implemented Patent Cooperation Treaty wherein a single standardized application may be filed in the Office which will be recognized as a regular national application in as many member countries to the Treaty as the applicant desires. This serves as an alternative to the present method of filing applications in each of the member countries. Two recently adopted procedures permit a reissue application to be examined in view of prior art submitted by a patentee or assignee which was not considered in the original examination, without amendment of the claims as filed, and a Protest program which permits any person to inspect a reissue application and submit prior art not previously considered. A most significant change in the Patent and Trademark Office would be occasioned by Congress passing a bill currently pending in the Senate which would make it an independent agency which would permit the Commissioner to make his budget requests directly to Congress.

SUMMARY

Our patent system evolved from methods of encouraging and recognizing invention to promote an expanding technological base during the colonial period. Such encouragement was accomplished mainly by societies for promoting manufacture, rather than by governmental action. The people who founded the United States were well aware of the importance of encouraging invention, and by establishing a patent system made such encouragement a governmental function. The patent system was changed and modified to keep pace with expanding technology.

A greater technological base existed during the colonial period than is generally recognized. The early inventors of agricultural equipment find their counterparts in the mechanics, blacksmiths and farmers of the colonial period. As the farm equipment field became more industrialized and the role of the engineer as an inventor assumed greater importance, the mechanics, blacksmiths and farmers continued to contribute inventions.

As the patent system has changed in the past, it is changing today to better perform its function.

REFERENCES

1. Brooke Hindle, "Emulation and Invention." The Anson G. Phelps Lectureship on Early American History originally delivered at New York University, to be published in expanded form in late 1980.

2. Brooke Hindle, "The Pursuit of Science in Revolutionary America." Chapel Hill, N.C., University of North Carolina Press, 1956.

3. Brooke Hindle, "Technology in Early America." Chapel Hill, N.C., University of North Carolina Press, 1966.

4. Bernard Jaffee, "Men of Science in America." New York, N.Y., Simon and Schuster, 1944.

5. Mitchell Wilson, "American Science and Invention." New York, N.Y., Simon and Schuster, 1954.

6. Raymond P. Stearns, "Science in the British Colonies in America." Urbana, Ill., University of Illinois Press, 1970.

7. Carl Van Dorn, "Benjamin Franklin." New York, N.Y., The Viking Press, 1938.

8. Saul K. Padover, "The Washington Papers." New York, N.Y., Harper and Brothers, 1955.

9. "The Story of the American Patent System 1790-1940." Washington, D.C., 1940.

10. "The U.S. Patent Office Since 1790." Washington, D.C., 1901.

11. John F. Steward, "The Reapers." New York, N.Y., Greenburg, 1909.

12. R. B. Swift, "Who Invented the Reapers?" 1897.

13. Edward Stabler, "Overlooked Reaper History." Baltimore, Md., Mills and Cox, 1854.

14. Herbert Casson, "Cyrus Hall McCormick." Chicago, Ill., McClurg and Co., 1909.

15. "Lincoln Lore." No. 1516, The Lincoln National Life Insurance Company, Fort Wayne, Ind., June 1964.

16. Stewart H. Holbrook, "Machines of Plenty." New York, N.Y., The Macmillan Company, 1955.

17. Paul C. Johnson, "Farm Inventions in the Making of America." Wallace-Homstead Book Company, 1976.

18. "Annual Report of the Commissioner of Patents." 1839-60 and 1870-75.

800940

Your Patent System

William A. Murray
Deere & Co.

WHERE DID THE U. S. Patent system begin. Simply stated, it began with our Constitution, written in Philadelphia in 1787, Article 1, Section 8, which provides in part "that Congress shall have the power...to promote the progress of science and useful arts, by securing for limited time to authors and inventors the exclusive right to their respective writings and discoveries". In our day-to-day association with patents, our primary concerns are whether there is infringement or not infringement or whether a particular patent is valid or invalid. However, the writers of the Constitution, when they prepared Article 1, Section 8, had in mind "...to promote the progress of science and useful arts". I would like to raise the question as to whether, in fact, the patent powers that Congress has provided under the Constitution, do "promote the progress of science and useful arts"?

As we sit here today, we are illuminated by electric lights. I mention this because 100 years ago, Thomas Alva Edison invented the electric light. He obtained a patent on the electric light and was granted the so-called patent monopoly which was the right to prevent others, over a period of time, from using or infringing his patent. The question is whether it did promote the progress of science and useful arts. Actually, Edison made very little money on the electric light bulb. By the time electric wires were strung in sufficient quantities to create a market for his product, his patent had almost expired. However, as a result of Edison's invention, power plants were built, electric wiring was extended to and between houses and factories, primarily for the purpose of giving light inside of the respective buildings. I recall one of the first homes that I lived in was a home built in about 1880 and there were still provisions for gas lighting along the walls, even though we had electric lighting. I might also add that the wiring was strung along rows of porcelain insulators, one row carrying the black, or positive wire, and several inches away, a row carrying the white, or negative wire. However, with that wiring, which was undoubtedly installed in the house for the electric lighting, we were able to operate toasters, electric clocks, mix masters, radios, and eventually television. The question that comes to my mind is whether had Edison, or someone else, not invented the electric light, wiring would have been placed in the homes to operate these accessories. I question whether the advance or promotion of the associated electrical arts would have been developed at the early stages that did occur. In otherwords, the promotion of the art, as developed by Edison, put electric wiring in homes. Once the

ABSTRACT

Today our primary concerns regarding patents are whether there is infringement or not infringement or whether a patent is valid or invalid. The writers of the Constitution, however, had this in mind " . . . to promote the progress of science and useful arts".

The material presented deals with the question as to whether, in fact, the patent powers provided by Congress under the Constitution do "promote the progress of science and useful arts".

SAE/SP-80/470/$02.50

electric wiring was there, engineers, such as the early counterparts to yourselves, quickly said, "Hey we've got electric wiring, let's utilize it for other purposes. Any place an electric motor is needed, we can use that electric wiring. We can use it, for example, to operate the resistors in toasters. We can use it to drive the motors for mix masters. We can operate it to run the electrical circuits on radios." Promote the useful arts? A perfect example.

When I was much younger than I am now, the nation, or at least the government, was quite concerned about the power of Eastman Kodak and its control of the market for films, as well as cameras. Historically, Eastman Kodak, as it improved its film and cameras, obtained patent protection and, to a degree, used the patent laws to its great benefit. However, Thomas Edison, using the principles of Eastman photography, developed the motion picture projector. Further, in 1948, a fellow named Land, building on the art developed by Eastman Kodak, came out with a camera that developed film immediately after taking pictures. The United States Patent Office granted him a patent. As a result, Eastman Kodak has had its greatest competition for the American market. The original patents obtained by Land have now expired. However, the monopoly that was granted to Mr. Land to exclude others for the period of 17 years gave him the opportunity to develop a market, to research and further develop his specific field, and to produce cameras and film for his Polaroid cameras. On the basic premise that a picture could be taken, a company such as Xerox developed machines that would take and develop pictures instantaneously. It, too, obtained several patents which gave it the opportunity to develop and sell its product and to build an industry around this product. Therefore, did Eastman Kodak, in its original patents and the protection that was granted to it, promote the useful arts? Certainly, it did.

The above are all great success stories relating to development of products that have had an impact on all of us. However, the patent system promotes success in smaller quantities. About 17 years ago, a young farmer approached Deere & Company with an idea concerning a manure spreader. Manure was always a dirty subject and consequently, the farmers wished there was something they could do to improve the handling of it. This young farmer had the idea to make an upwardly opening semi-cylindrical tub, closed at both ends, for carrying manure. He mounted a fore-and-aft extending shaft with chains mounted on the shaft lengthwise of the tub. When the tub was filled with manure, rotation of the shaft caused the chains to wrap around the shaft and beat a path through the manure which would, in effect, discharge it over one side of the tub. Deere & Company purchased this idea for the reason that it felt that there was a market for a spreader that could handle liquid type manures which it believed this invention could best handle. Patents were applied for. Fortunately, for this young farmer, Deere later examined the market and felt that there was not a sufficiently large market to warrant its manufacturing this new manure spreader.

The young farmer farmed outside a small town in Iowa. The town was, like so many other towns in Iowa, losing its citizens and the farmers in the area to the warmer climates of California, and primarily because the farms were becoming larger and required fewer services and less manpower. Several of the businessmen approached the young inventor farmer and proposed that they invest in and start their own industry in the town based upon his patented spreader. The manure spreader became successful. It did, in fact, create the largest industry in this small town. It employed several hundred people from the small town and the surrounding farm area. As other companies approached the small company and asked for a license, they were refused. Some companies threatened that they were going to disregard the patent. However, none did. For 17 years, the small factory was able to establish a market, expand its business to other lines, build up a marketing and sales group, and become a subsidiary of a relatively large conglomerate. It was successful. Why? Because it had a small piece of paper called a patent which said to other companies, "If you infringe this patent, we will sue you." Did it, therefore, promote the useful arts as our forefathers desired to grant in their patent granted period? You bet it did.

If you keep in mind that the purpose of the patent influence in our Constitution is to "...promote the useful arts", you should become less concerned about a phenomenon which has occurred in the last few years, which is the tremendous influx of foreign owned patents that are obtained in the United States Patent Office. If we appreciate that the purpose of the patent system was and is "...to promote the useful arts", we should not be alarmed where the promotion of the useful arts comes from, whether it be from Germany, Japan, Russia, or others, so long as it promotes. This probably doesn't satisfy your day-to-day problem of, for example, when, in a development program you are faced with a Japanese owned U.S. patent which blocks your path. If it is a good patent, you take a license. If it is a poor patent, you ignore it. If it is a "maybe" patent, you consider taking a license. So many people consider 17 years to be such a long time for which the patent monopoly is granted. In almost all instances, the first portion of that 17 years goes into a development stage in which no profit is shown the inventor. There is a short period, normally, where the inventor does have a monopoly to develop his product, to develop markets, and if possible, to secure sufficient manufacturing facilities. However, from a practical standpoint, most patents, in the present days, do not have a 17 year practical life. There is constant improvement and leapfrogging in industry which, in effect, most often makes a patentable improvement obsolete after it is on the market four or five years.

On a different aspect of patents, I believe there is quite a bit of mystery to engineers and inventors, such as yourselves, as to what constitutes patentable inventions. However, I believe there are some fundamentals that you can learn that will get you through day-to-day thinking regarding your improvements or inventions.

Everyone invents and continuously invents. For example, some years ago, when women used hairpins, it was known that a woman could do several things with a hairpin. She could certainly tighten screws with them, she could often clean a drain with them, she could repair the gas stove with them, she could unlock doors and cabinets with them and basically, it was known that to give a woman a hairpin, there were very few things she couldn't do. These uses of a hairpin were, in fact, inventions. If she used the hairpin as a paper clip, it was a new use for a hairpin and she, in a small way, had created an invention. Certainly, the secretaries in your office have invented a new way of filing and certainly, you have invented your own way of filing your papers on your desk. For example, on my desk, I have one pile of papers that I've got to do right away, a second pile, which is less important because I didn't do them yesterday, and a third pile which is generally "will do when I get time" because I should have done them a few weeks ago. To a degree, I have invented my own filing system.

The question then becomes, "What is a patentable invention?" For a long time, in the legal profession, we spoke of invention and not patentable invention. An invention consisted, at one time, to be that which had a use and which was new. This led to all kinds of problems for, after all, a four-legged stool should not be an invention over a three-legged stool even though it would be new as compared to a three-legged stool. A porcelain doorknob should not be considered an invention over a metal doorknob even though it was new and useful. Justice Douglas, in some decisions written by him, was one of the first to insist that something additional other than the invention being new and useful should be required to make it patentable. In a decision prior to 1952, the United States Supreme Court set forth a third requirement for a patentable invention that there must "...be revealed a flash of genius" in the creative process.

This, of course, created even more problems, for who could tell when a "flash of genius" occurred. Attorneys were arguing that "a flash of genius" did or did not occur, depending upon the respective side they were representing. The patent system was, therefore, at the time, in somewhat disarray. It was necessary that Congress take some action to alleviate the conditions. The U. S. Congress passed the first comprehensive patent statute which set forth three criteria for a patentable invention. To be a patentable invention, the invention had to be new, useful, and, it had to be such that the differences between the subject matter sought to be patented and the prior art are such that the subject matter of the invention, as a whole, would not be obvious, at the time the invention was made, to a person having ordinary skill in the art. In order to get rid of the "flash of genius" test, the statute provided "Patentability shall not be negatived by the manner in which the invention was made." Therefore, as to whether you have a patentable invention gets down to the basic question of whether your improvement would have been obvious to one skilled in the art at the time it was made. More simply, the question of "patentability" is a question of "obviousness".

It can be said that "obviousness" is certainly a judgment decision on your part or on the part of a court. It is like determining what an ordinary man would do under certain conditions. However, the U. S. Supreme Court, in 1966, took a very hard look at the 1952 statute and did set up a criteria for determining "obviousness", and while I recognize that I may be over-simplifying the entire question, I think it did leave a basis from which a group such as this can work and, as they say, leave the tougher questions to your patent attorney. The Supreme Court, in 1966, in Graham vs. John Deere, set forth the basis for determining whether an invention was obvious to one skilled in the art. It is the leading case, even today, with respect to the issue of obviousness. It is cited in more lawsuits regarding patents and there are probably very few cases, since the decision, wherein a question of validity arises that Graham isn't cited. Therefore, if you understand Graham vs. John Deere, you will get some idea as to the criteria for obviousness.

A little background on this case may be in order. Graham had invented a clamp for mounting a chisel plow shank on a tool bar. The clamp had a spring device which permitted the chisel plow to rise and fall in accordance with the characteristics of the ground. For example, if it encountered a stone, the spring would permit it to yield so that it would not break the shank. Graham had had considerable success with this clamp. He had previously sued another infringer. The patent was held valid and infringed both in the district court and the court of appeal. Graham then sued John Deere in a different federal circuit. In the suit against Deere, in the lower court, the patent was again held valid and infringed. However, upon appeal, the Court of Appeals, in that circuit, held the patent to be not valid. This, of course, raised the red flag to the U. S. Supreme Court. As you may be aware, the Supreme Court desires to have every federal circuit apply the same law of the land. When one circuit is in conflict with another circuit, often the Supreme Court will take the case and decide which circuit has applied the law correctly.

The historical significance of the Graham vs. John Deere was that it was basically the first case the Supreme Court took up on the issue of obviousness. It was recognized that it was going to be a landmark decision. Briefs were filed on behalf of not only the plaintiff and defendant, but on behalf of many of the state bars and the American Bar Association, setting forth how the respective parties believed the 1952 statute, relating to obviousness, should be applied. Consequently, the U. S. Supreme Court hit the question head-on and did decide how obviousness should be determined. The Supreme Court held that the manner of determining obviousness would be to collect the best known prior art and determine the differences between that prior art and that desired to be patented. Obviousness would be determined by deciding whether those differences were obvious to one ordinarily skilled in the art.

The Supreme Court treated as secondary evidence proof of commercial success, long-felt need, or failure by others to succeed in determining the question of patentability. The court was clear to point out that if the invention desired to be patented was obvious to one skilled in the art, that no amount of commercial success, long-felt need, and failure of others would change it to a patentable invention.

In closing, I would leave one further thought to you as engineers and people in the development areas of your respective industries. In determining whether you have a patentable invention, remember that the issue is whether, at the time of the invention, it would have been obvious to one skilled in the art. I would like you to reflect just a moment on the portion of the statute "... at the time of the invention." What I would recommend to each of you is that as you make improvements in your respective products, you do not minimize those improvements. Engineers, by their nature and training, are inquisitive with respect to how a product is constructed and operates. You engineers must completely understand the functions, purposes and operation of the various parts of your products. This leads, in my opinion, to an over-simplification of your own advancement or improvements. There is only a step-by-step progression from "I understand it", to "Any good engineer can understand it", to "Any good engineer could have developed it", to the ultimate conclusion "It would have been obvious to any good engineer". There is a falicy in such thinking which is that you start out with an accomplished fact (which is your improvement) and work back. Don't forget that at the time of making your improvement, this accomplished fact did not exist and from the state of the art, you worked forward to accomplish your improvement. Remember Thomas Edison merely took old and well-known elements to create his electric light. He took old and well-known film and mechanical products to create a motion picture projector. In otherwords, he took old and well-known products from different shelves, put them together and made inventions...and they were patentable inventions. I submit to you that if you take old and well-known products and put them together to create a product that was not in existance before, don't look at that final product and say "Anyone could have done it". I would recommend to you that if you create a new product that gives any kind of an improved function, regardless of whether each of the elements that you have combined are old or not, don't mentally look at the end result with the hindsight of that end result itself. Often, there is a very close line between that which is obvious at the time of the invention and that which is obvious with the hindsight of the invention. More often than you may think, there isn't a fine line, but a wide gap. The mere fact that you are using and combining old elements does not prevent your having a patentable invention. It must be an obvious combination at the time of your invention...not the hindsight required after having viewed your invention.

800941

Safety Accomplishments on Agricultural Equipment

Carlton Zink
Product Safety Consultant

BORN ABOUT 8 O'CLOCK one Saturday morning in May, 1902, in a farm house in eastern Nebraska, I grew up in the early days of farm tractors.

In our own immediate family of father, mother, and four children, there were no tragic accidents. On a wall in our living room, there was a painted photograph of a small boy's head and shoulders. This, I learned, was of my mother's little brother who died before she was born. Not until I was older, perhaps in my teens, did I learn the cause of his death. He unexpectedly came around the corner of my grandfather's stable as he was cleaning a horse stall and the boy was impaled on the swinging pitchfork.

The fork is a simple tool, but one difficult if not impossible to guard effectively except as the users exercise care. Other hand tools, such as the axe, the sledge, even the housewife's "butcher knife" are all dependent on the individual for their safe use.

Horses, mules, oxen, all were used in those early days as draft animals, but there were also ways to use their strength to meet mechanical needs.

If the amount of power required was not great, one might see a horse or two on a treadmill. They walked and walked, without moving forward, on an endless slatted floor which as it turned operated some small grinder, thresher, or farmstead machine. A goat or a large dog might power the family butter churn.

For larger machines one might see four teams (eight horses) going round and round, each team hitched to a radial beam. These beams were spaced 90 degrees and fastened at the center to a large gear. Other gears meshing with the large one applied power to a tumbling rod running out at ground level and terminating in a pulley and belt, sprocket and chain, or gears and so supply power to one of the larger machines.

My father, in the 1890s, tended the separator or thresher on such an outfit. We were speaking of tumbling rods. Occasionally, an individual wearing work pants with frayed bottoms would step over or close to the tumbling rod and the pant leg would catch. If the cloth was old it might break loose, but if strong, it could wind around the rod and cause serious injury. Sometimes a loud "Whoa!" could bring relief, but not on the modern counterpart, the tractor PTO as many have discovered. A couple of boards nailed together to form a trough were sometimes placed over the tumbling rod in an inverted vee fashion. There may have been some commercial shields, but I never saw any.

In the first 15 to 20 years of this century steam power was popular. Threshing, shelling, clover hulling, shredding, some plowing, all were done with steam power. I loved the steam engines and would sit on them

ABSTRACT

The material presented deals with safety accomplishments on agricultural equipment.

An introduction on the early days of agricultural equipment is presented including the development of the power take-off.

A complete list of hazardous situations, areas, or mechanisms associated with agricultural equipment is provided along with discussions on what has been done to reduce or eliminate the hazard.

Finally the appendices cite SAE and ASAE Safety Recommendations and Standards.

SAE/SP-80/470/$02.50

at every opportunity. Carrying drinking water to the threshing crew was a job for a young teenager. This provided occasional moments when one could be around the engine or thresher.

Hazards, yes, there were some. The band cutters over the grain bundle feed table just in front of the threshing cylinder were wicked. Pitch forks were sometimes inadvertently thrown into the moving knives, but I never saw or heard of a person being caught in them.

All of the belts on the side of the separator were exposed as was the main drive belt from engine to separator. I know of one instance where the inside horse on a "bundle wagon" team flicked its tail up and into the pinch point of the main drive belt and pulley of the separator and had his tail pulled out by the roots.

Running so low on water that the steam engine blew up did sometimes happen. The engineer watched the water level carefully and more than once I have heard a series of whistle blasts which was a signal to the "water boy" to hurry with his tank load of water for the boiler. The steamers were replaced by large, heavy tractors. Some were built on a steamer-like chassis. Most were large and cumbersome - few were easy to start.

The Moline Universal, built in the late 'teens was, I believe the first tractor to have battery ignition and a high speed engine (1800 rpm). It is interesting to note that International's row crop tractors are built on the factory site used by the Moline Universal some 60 years ago. Some changes have been made, I am sure. When McCormick began development of the reaper and with others carried the concept along into binder and mower, they used a cutting mechanism, the sickle, that has persisted through the years. The fact that the sickle must operate in an unguarded condition to be effective has resulted in injury to some wildlife and an occasional child hidden from sight in the crop being harvested. The Oregon legislature once had a bill under consideration requiring that cutter bars should be shielded. It must not have been introduced by a farmer.

In the 'teens and 'twenties other crop harvesting or processing units were introduced. The silo filler (a cutter and blower) was one, the mechanical corn picker was another. In operating a silo filler, corn bundles were placed on a conveyor type feed table. This endless conveyor carried the bundles forward to where the stalks were seized between two large toothed rollers. These rollers carried the bundles through to the point where a multi-bladed revolving knife cut the stalks into small bits.

A hazard existed for the individual who might stand alongside the feed table and facilitate the entry of the bundles. He might retain hold of a bundle a trifle too long and be caught in the rollers - the "lion's mouth" as I have termed it.

A clutch that could be quickly released was seen to be desirable and if not initially, was soon incorporated into the design. To ensure quick action the throw-out lever was usually an arched rod coming up from each side of the feed table, high enough that it would not interfere with the bundles, but would stop the feed table and rollers with hand or body pressure against it. This clutch handle was often referred to as the "monkey bar" or "trapeze bar".

The snapping rolls on a corn picker accounted for many injuries, often a hand, sometimes a leg, occasionally fatal. The rolls would plug - stalks would not go through - warning decals told the operator to shut off the machine or at least disconnect power to the rolls before unclogging. Users considered this an easier task if the rolls were turning. Tugging at the stalks sometimes caused them to move. The stalk then usually moved at such speed that one could not get a message from brain to hand to let loose of the stalk - and the hand would be in the rolls before it could be snatched back.

Although the problem of providing quick relief for a man caught in the rolls has been widely discussed, I have never seen or heard of an effective de-clutching means for the rolls and/or gathering chains that would be accessible to the entrapped individual. Similar gathering devices have been used on another popular row crop machine, the field forage harvester. Not as many injuries have occurred as with the corn picker. De-clutching devices if at all available, are not common. I know of one instance where the operator was feeding material, pulled out of a clogged machine, back into the rolls. One leg became caught in a loop of hegari (tall cane). He came to rest with one leg in the feed wagon and the feed rolls ("the lion's mouth") chewing on his pelvis. He lived and when I saw him, he was moving about using one crutch. Several unsuccessful attempts were made to adapt a prosthetic to his situation. The corn head used on combines when they are harvesting that crop still must operate unguarded. The use of "stripper bars" above the rolls reduces the hazard to some degree.

Machine capacity adequate for the crop is essential. I once heard an engineer in charge of corn picker design say it this way: "Don't try to harvest a 125 bushel crop with a 75 bushel machine."

To this point, we have discussed hand tools and some harvesting equipment, most of which must function without guarding devices that would protect the operator against any inadvertent entanglement.

THE POWER TAKE-OFF

Early harvesting machines requiring power were driven by a ground wheel, usually called a "bull wheel". If the harvest fields were quite muddy, lack of traction was a real problem. Small engines were mounted on some machines, permitting harvest to go on even if the binders were pulled on a sled or "stone boat". This was particularly true due to heavy rain in 1914 and 1915 in the Dakotas, Nebraska, and Kansas. Many engines mounted on binders for harvest time were later used to pump water for livestock.

Implement engineers were increasingly impressed with the need for some means of using surplus tractor power to operate the drawn machines. Out of this came the power take-off, a shaft coming out of the tractor transmission which carried the power back to the drawn implement. International Harvester is given credit for early use of the PTO to power a grain binder. This was in the early to mid-'twenties. There has been slow but relatively continuous improvement in the safety situation on the power take-off even to the present date.

This connecting shaft - tractor to machine - was telescoping to accomodate movement over ridges and through swales and had a universal joint at each end to take care of variations in alignment. There is evidence of early recognition of the entanglement hazard. In the 1930s shielding appeared. These devices consisted of two over-hanging shields, one fastened to the tractor and the other to the implement with their outer ends almost but not quite touching.

The next plan involved a telescoping U-shaped sheet metal shield, with top and sides, attached at one end to the tractor and at the other to the implement. The attaching points at both tractor and implement were short shields covering the exposed shaft on the tractor and the outer end of the universal joint on the machine. When in operation the front half of the front joint was under the tractor "master shield". Because of need for clearance, the shielding was less than complete and injuries occurred when pant legs, shoe laces, and the like were caught, particularly on cotter keys, latching pins, or other projections on the rapidly revolving front universal joint. Many users were not impressed with the desirability of having the shields in place whenever the tractor and implement were being used. The exposed PTO shaft was a still greater hazard and many injuries could be charged to it. A quite common user attitude has been: "It can't happen to me."

To discourage removal of the inverted U-shaped shielding, a flat strap or "belly band" was welded to the underside of the shield. This may have helped, but some farmers, and dealers too, removed the straps with chisel or cutting torch.

In the early 1940s Herman Altgeldt of Oliver at South Bend and George Harrington of Blood Bros. at Allegan, Michigan combined efforts to develop and produce a telescoping, spinner type shield for the PTO. Bell-shaped stampings were welded to one end of each tube and the outer ends of the tubes were journaled on the inner half of each U-joint. The widely spaced balls in the bearing together with its vulnerability to sand and dirt resulted in some early failures and removal of the tubing rather than replacement of the bearings. Though quite effective when in good condition, they were not widely adopted.

In the mid-1950s Roy Harrington of Deere developed a successful nylon-strip bearing for the spinner shield. By the late 1950s the industry had replaced the inverted-U sheet metal shield with the nylon bearing spinner type. The ball shaped ends on the telescoping spinner cover the inner ends of the U-joints and the outer ends are under the tractor or machine master shields. Although this combination is still vulnerable to entanglement on the bottom and sides of the master shields, it is widely used in current production.

In 1966 the specifications for a fully shielded power take-off power line for both towed and integral machines were adopted as SAE J955 and ASAE S297. In December, 1968, it was classified as a tentative ASAE Standard, reconfirmed each December through 1975, revised March 1977, reconfirmed December 1977 and 1978. SAE last published J955 in 1974. Deere & Company developed this concept, suppliers cooperated and Deere uses this full enclosure quite extensively.

Walterscheid of Germany, for a number of years, has made a relatively full enclosure unit that was restrained from turning by a short chain - the power shaft revolved within the enclosure.

Roughly 60 years has passed since the introduction of the power take-off - slow but consistent progress has gone from a totally unshielded situation to the development of, but limited use of, a full enclosure.

VARIOUS HAZARDOUS SITUATIONS

Let us assemble a reasonably complete list of hazardous situations, areas, or mechanisms associated with agricultural - field and farmstead - equipment. We shall list them and discuss in turn what has been done to reduce or eliminate the hazard.

1. ENTANGLEMENT - in gears, chains, belts, augers, feed rolls, revolving shafts (smooth or with collars), protruding bolts, cotter keys, and the like.
2. OVERTURNS - of tractors or implements.
3. OPERATOR CONTROLS - Confusion due to location, required direction of movement, inadequate design leading to unexpected - sometimes disastrous results.
4. OPERATIONAL AND SERVICING AREAS
5. BRAKING AND PARKING
6. HIGHWAY TRAVEL - collisions, overturns. Desirability for lighting and marking - identification of slow-moving vehicles.
7. FIRE
8. OPERATOR'S MANUALS, SAFETY INSTRUCTIONS, AND DECALS

Entanglement - Clothing and body parts may become caught in, on or around some powered, moving mechanism causing injury, often serious, sometimes fatal. Earlier we discussed the power take-off and showed that the situation has moved from the totally unshielded to a relatively, but not entirely, safe condition for most currently produced machines. We also showed that a still safer design exists with somewhat limited application.

Corn pickers and corn heads on combines, have gathering chains and snapping rolls that must operate unguarded to function properly. Forage harvesters with row crop attachments use a gathering device, sometimes a chain with long encompassing "tongues", on others a series of interlocking belt folds which grasp the crop stalk. A power shut-off within probable reach of the victim that would still permit free flow of the crop seems possible; I have seen none to date.

There are many potential sources of entanglement that could be so shielded that there would be no adverse effect on machine function. Gears, chains, and belts can often be adequately shielded at their "nip points". In some instances one large shield may be used to cover a nest or cluster of the gears, chains, or belts. It may be wise to hinge such shielding so that it will remain attached but can readily be swung aside for servicing or adjusting the machine, then closed and secured.

All of the entangling mechanisms mentioned may, in some designs, be sufficiently shielded by their location. This is interpreted to mean that "inadvertent contact is minimized during normal operation or servicing".

Smooth revolving shafts are sometimes regarded as being relatively harmless, but in my limited experience, I have known of two individuals who lost arms because of jacket sleeves catching on and wrapping around a smooth shaft. Connecting collars and projecting bolts appreciably increase the degree of hazard. Metal or plastic tubes over the revolving shaft are sometimes appropriate. If a shaft is located against some portion of the machine, then a metal cover for the other 270 degrees will provide protection. In some situations, metal tubes will rust or be filled with foreign material and revolve with the shaft; plastic tubes may deteriorate, break up and leave the shaft exposed. Such situations throw an added burden on the dealer's service and parts departments.

Augers, chain and slats, and wide belts may be used to convey material. The augers present the greatest hazard of the three means. The revolving auger usually operates against a cut-off point. The force exerted has often proved sufficient to snip off hands, arms or feet, and legs. I have not seen any exposed augers that could not have been adequately shielded at nominal cost except those which must be open to accomplish intended function as for example on a combine header.

Auger elevators, which for a number of years after their initial production in the 1950s, used exposed auger intakes now are equipped with a standard protective screen opening believed to be optimum for grain movement while still affording protection against inadvertent contact.

Overturns - Overturns took place rather infrequently when tractors first appeared in the 'teens and 'twenties. Usually in those early years they overturned to the rear due to hitching too high, trying to pull some immovable object such as a large stump, or being bogged down and if power was still applied the front end was cranked over backwards. The low horsepower to weight relationship in early tractors reduced the chances of rearward overturns.

The introduction of rubber tires for tractors in the mid-'thirties resulted in faster, more versatile tractors. The owners of the light, nimble Ford-Ferguson used it for various "cow pony" jobs, in many instances over rough terrain. There were upsets to the side, as well as to the rear, and occasionally to the front. Tricycle row crop tractors overturned to the side more readily than the conventional four wheel standard type. Used on the highway to pull grain wagons or transport implements, it was not unusual for them to overturn to the side into a roadside ditch. Rough berms, washes, chuckholes, and the like sometimes caused upsets. If a tractor front wheel fell off of the edge of the road, many drivers would try to scramble back on. This sharp movement would hasten the upset. A hard quick turn into the ditch would have, in many cases, kept the tractor right side up. In the early to mid-1950s some work was done at the Davis branch of the University of California, by Agricultural Engineering Department members and a cooperating tractor company, to develop overturn protection. Though the results had merit, little enthusiasm was evidenced by users or manufacturers. In this same general period Sweden developed roll-over protection and made its use mandatory. Interest increased in the United States and in the late 1950s and early 1960s various groups, universities, and manufacturers began efforts to build and try out protective structures. Two states, North Dakota and Illinois, designed and built protection for the operators of their highway mowing tractors. Interest and activity was heightened by the accumulation of data that indicated that well over 50% of tractor fatalities were due to upsets. In March 1966 at a cooperative meeting, sponsored by ASAE and FIEI, the overturn problem and how to meet it was thoroughly discussed.

In response to a statement by the Illinois State Highway representative that they had not been able to purchase overturn protection, Merlin Hansen of Deere told him that such units would be available in September. Deere and International Harvester led the way with other manufacturers following closely. Overturn demonstrations at the University of Nebraska Power and Safety Day for three years, the development of movies by Deere and International Harvester, together with support by other farm safety minded persons, stepped up enthusiasm for this protection. Prospective purchasers, looking at the cost of the "roll bar and canopy" decided in increasing numbers to purchase a cab, reinforced to give adequate protection but with relief from bad weather and dust. The FIEI Engineering Committee developed a proposal for tractor operator protection and for construction and testing of roll bars and of cabs. Initially this was adopted and covered by SAE J333 and J334 and by ASAE R305 and R306. Shortly SAE J167 and ASAE S310 were added, followed by SAE J168 and ASAE S336. In 1977 several of the earlier standards were replaced by SAE J1194 and ASAE S383 ROPS for wheeled agricultural tractors. Production of ROPS for earlier tractors has been discouraged, these tractors were not designed to withstand the loads that would be imposed by overturns.

Operator Controls - Operator controls on early tractors varied in location and direction of operation according to the ideas of the tractor designers. This could and did lead to mishandling, particularly when one individual might be using two or three makes from day to day. In 1962 proposed standardization of controls appeared in ASAE R234 - Tractors and R235 - Self-Propelled Implements and for SAE in J841. These were superseded by the current standard ASAE S335.1 - Operator Controls on Agricultural Equipment. SAE J841 was continued and ANSI B114.7 applies.

Since many tractor makes and models are widely distributed over the world, there is need for universal symbols that will identify on any tractor, in any country, the function of each operating control, the meaning of the readings of the various instruments, as well as the need and points for servicing the tractor or implement. In 1967, ASAE Standard S304 - Symbols for Operator Controls on Agricultural Equipment was adopted and with revisions is now current. These symbols are also covered by SAE J389 and by ANSI B114.2

Operational and Servicing Areas - Falls have accounted for many of the injuries incurred when the individual was on the operator's platform and when servicing the larger machines such as a combine. Consideration of ways and means to bring a greater degree of safety into these areas began in the mid-1950s. ASAE R275 on Implements and R280 on Tractors first carried in the 1965 ASAE Yearbook each had a section on operator work positions. Slip resistant surfaces on platforms, steps, walkways and hand holds, guard rails, or other safeguards were suggested. Access ladders with front foot support and shielding where necessary to keep toes out of moving parts was recommended. Glazing material, as might be used in cab windows, was to be of a type and quality that would not fragment into dagger-like pieces.

These two recommendations were merged into our now current ASAE S318 - Safety for Agricultural Equipment. SAE J208 and ANSI B114.3 cover the same area of concern.

Braking and Parking - The means to reduce a tractor's free-rolling speed and to ensure the tractor remaining where one stopped it, particularly on a grade became more important with the increased use of rubber tires.

An early compelling need for brakes was on the first row crop tractors. The operator would make a 180 degree turn around at the end of each pass through the field. The early IH Farmall actuated the steering brakes by a sharp right or left turn of the steering wheel. Other tractors used right and left foot pedals. These early brakes were not consistently effective, their lasting qualities were not noteworthy. Many a section of fence was removed by hooking into it with the outside front corner of the cultivator. Brake improvements came through the years, one popular brake by a supplier was a dry disc type, actuated by a steel ball moving on a ramp. Later hydraulic brakes were used, and are still current. Early parking brakes were usually external contracting band type actuated by a hand lever, held by a pawl. Locking two transmission speeds together with the gear shifter has been an automotive practice for many years and is used on some current tractors. ASAE S365T - Brake Test Procedures and Brake Performance Criteria for Agricultural Equipment covers the needs in braking and parking.

Highway Travel - Highway travel of tractors and agricultural implements has increased greatly over the years. The wholehearted acceptance of rubber tires made this a common practice. The economic advantages of farming larger acreages caused farmers to lease or buy farms, perhaps even several miles away from the home farm. This put more tractors and implements on the public roads, often after nightfall.

In November, 1953, the FEI Lighting Subcommittee of the Advisory Engineering Committee developed a safety lighting program. It was submitted to ASAE and adopted by ASAE as a Standard in February, 1954. Not numbered at that time, this standard was known as Safety Lighting for Combinations of Farm Tractors and Implements and appeared on page 51 of the 1954 ASAE Yearbook (the first year it was published).

This standard called for taillights or red reflectors and a safety lamp with red to the rear and amber to the front, powered from the tractor. As of 1965, this standard, with revision, appeared as ASAE S279. SAE dealt with this in J137 and ANSI in B114.6.

In the discussions relative to the movement of agricultural equipment on the highways, it became evident that some, perhaps many accidents were caused by the driver of the over-taking vehicle mistakenly judging the unit up ahead to be moving nearly as fast as he might be. Finding out too late that it was moving slowly the "over-taker" might not get around and past safely.

A search for an effective identification for a slow moving vehicle or combination was headed up by Ken Harkness, a graduate student in Agricultural Engineering at Ohio State with a grant from the Automotive Safety Foundation. Minnesota Mining and Manufacturing, makers of reflective film, cooperated with Harkness. With reflective material supplied by Harkness, Deere made up nearly 100 slow moving emblems, supplied them to other manufacturers and safety specialists in farm organizations and universities. These persons tried the emblem and fed back their findings. An ASAE Standard, S276 and SAE J943 were developed cooperatively by the ASAE Farm Safety Committee and the National Safety Council's Studies and Research Committee. ANSI's B114.1 deals with the same subject. The current lighting and marking has become much more sophisticated and effective, but the SMV emblem is still valuable for its intended purpose, that of telling the oncoming driver that the unit up ahead is moving 25 miles per hour or less.

SAE J208, ANSI B114.3, and ASAE S318 all call for such other safety items as:

1. at least one rear view mirror if operator is in an enclosure;

2. "hitch pins and other hitching devices" shall have a retainer to prevent accidental unhitching;

3. components on a wide unit, such as a disk harrow, that retract for transport shall have means to secure them in a retracted position;

4. provision shall be made for the use of auxiliary attaching systems per ASAE Standard S338 - Safety Chain for Towed Equipment, on equipment which is towed on highways by single point attachment.

Fire - Fire could easily destroy a tractor or combine. In fact, any piece of agricultural equipment that has a fuel tank is vulnerable. On machines which are trash accumulators, "shields shall be provided for the engine exhaust manifolds, muffler, and exhaust pipe when necessary to prevent contact with flammable crop materials" according to Section 12, S318. Fuel sediment bowls should be fire resistant. Early common glass bowls would break if fire surrounded them and then the fuel tank would empty and feed the fire.

Fuel temperature may rise quite a few degrees when equipment is being used on a hot day. It is important that the fuel tank cap be vented. On some tractors the radiator cap (unvented) and the fuel tank cap (vented) are the same size and general appearance. Use of a non-vented cap on a fuel tank can be disastrous if pressure builds up and the tank cap is removed. A number of fires have resulted from such action.

Safety Signs - Safety signs are useful in warning the operator and others of potential danger "during normal operations and servicing".

SAE J208, ANSI B114.3 and S318 again cited, call for the use of safety signs. Alerting words could be Caution, Warning, Danger according to the degree of hazard.

For many years, the National Safety Council's Green Cross for Safety was used to alert operators to the importance of a safety message. NSC decided against industry use of their symbol. Industry's own Alert - an exclamation point with a triangle is covered by ASAE Standard S350 and SAE J1284. Certain colors are prescribed in this standard for the warning signs. The signs shall appear in the Operator's Manual. S318, J1208, and B114.3 further dictate the quality of material to be used in the decals and how they are to be tested.

Safety for Farmstead Equipment - Safety for farmstead equipment is treated in S354 adopted in 1972. Common knowledge, as early as 1950, applied to the whole range of agricultural equipment safety problems could have solved most of them - R275 and R280 grouped much of this knowledge in 1965. In the 1970s the Auger Elevator manufacturers and those concerned with Farmstead Equipment developed their own detailed standard S361T. Similarly, the Crop Dryer Manufacturers Council developed S248 which deals with the Construction and Rating of Equipment for Drying Farm Crops.

Another facet of farm safety is ASAE EP342 - Safety for Electrically Heated Livestock Waterers adopted in 1971. In the "early days" we simply built a retaining wall of 1 to 2 feet greater radius around the round stock tank and packed it full of horse manure. Others used a corn cob, coal, or wood burning watertight submerged heater; a popular one was known as the "cowboy" heater. Electric brooders for chicks, pigs, calves, lambs, and other similar stock are covered in ASAE EP258. In earlier years the mother of the young had to provide such comfort if there was any to be had. Not literally so, as the farmer would try to provide housing or shelter from severe cold.

Another popular machine on many farms is the rotary irrigation unit. It is often, but not necessarily powered by an electric motor. ASAE S362 tells how the wiring should be done and what the other important items of equipment may be.

The machine, properly installed, replaced the man with a shovel and that is assuming that ditch water was at the farm's boundary. The big rotary irrigation machines usually have a well close by and the pump may be engine or electric motor driven.

Lighting on farms has gone from the kerosene latern of my youth through the acetylene gas light, the Delco, and others, light plant to rural electrification. Large scale dairy and poultry operations have special lighting needs. They are covered in ASAE EP (Engineering Practice) 344. Essential warning signs are described and certain steps spelled out for installation and operation to reduce hazards to a minimum.

Provision is made in ASAE EP364 for Installation and Maintenance of Farm Standby

Electric Power. When the "high line" is out then "standby" power can assure continuity of farm operations. In the old days, the "standby" might well have been a candle or another lantern.

PROBLEMS PENDING

Conformity to existing standards is not complete. Neither do the operators always assume their share of the responsibility for safe operation. Designers sometimes overlook the fact that the better a machine functions, the greater the continuity of operation, the less often the operator is exposed to the hazard of correcting malfunction. A current example of the hazard attendant to malfunction has been the big, round baler. Since it first appeared on the market, operator efforts to correct malfunction have resulted in injuries bordering on epidemic proportions.

Tractor diesel engine starting and stopping controls could be profitably reviewed. Time may well correct safety problems, but for the person who may have been injured, time has run out.

I do not wish to close on a pessismistic note. I know there have been tremendous gains in the three-quarters of a century or more since the lifting of the burden from the farm worker's back began to accelerate.

Let us all, designers, manufacturers, sales and service persons, educators, safety specialists, all strive to hold these gains and add to them as the years, the men, and the machines go by.

APPENDIX A

The earliest safety standards or recommendations dealing with agricultural equipment that have come to light were published by the American Society of Agricultural Engineers. They are:

1. Recommended practice in the matter of Ratings and Specifications for Low Voltage Isolated Electric Light and Power Plants. This was adopted December 28, 1916.

2. The Selection, Installation and Operation of Farm Electric Motors. This was compiled and approved by the Committee on Electric Motors. Due to the increased interest and use of the electric motor by uninformed persons, it was felt desirable to publish information on selection, installation, and operation. This report was published in December, 1930.

3. Power Take-Off and Drawbar Hitch Locations for Agricultural Tractors and Machines. Initial efforts to develop a satisfactory PTO standard date back to December, 1926, with the issue in April, 1927. Revisions were adopted in 1928. In 1930 an intensive study was made covering principally strength of materials and shaft retaining means. The resulting standard was adopted in March, 1931. In 1940 and 1941 another intense study pointed toward the possibility of each attachment of any driven machine to any tractor resulted in the August, 1941 adoption of this revised standard.

A 1952 revision included a requirement that the tractor master shield should be strong enough to support the operator without taking a permanent set. The 1952 revision of the Power Take-Off Standard is the one that appeared in the first Agricultural Engineering Yearbook, published in 1954. A foreword under ASAE 203.9 - page 153 of the 1979 Yearbook tells in detail of the frequent revisions through the years from 1926 to 1978.

As shown on the chart in Appendix B, the Society of Automotive Engineers adopted like standards for a number of hazardous situations covered by ASAE. Some of them deal with the power take-off, lighting and marking of agricultural equipment transported on the public roads, the slow-moving vehicle identification emblem, control symbols that make it possible for operators, worldwide, to know what tractor control device to use.

Other mutual standards involve operator protection and how to properly test its adequacy, safety for all agricultural equipment and wheeled industrial equipment, and a common symbol to alert the operator to a potential hazard.

In Appendix B the various SAE and ASAE Recommendations, Standards, and Experimental Practices that included consideration of safety are listed. When they were first adopted, when and if they were discontinued, and a listing of those that are current are all covered in the chart.

Note that proposals which are initiated as recommendations may, with time and development become standards or experimental practices. The standards followed by the suffix T are tentative.

APPENDIX B

SAE AND ASAE SAFETY RECOMMENDATIONS AND STANDARDS

SAE	ASAE		
J1170	S203	POWER TAKE-OFF (540)	→ SEE S203
J719	S204	POWER TAKE-OFF (1000)	
		SAFETY LIGHTING	REPLACED BY S279
	EP258	INFRA-RED BROODERS	
J723(1)	S214	LIGHTING CONNECTOR	
J724(1)	S215	LIGHTING CABLE	
	EP256	MILK COOLER WIRING	
	S216	WARNING LIGHT - SELF-POWERED	
J841	R234	TRACTOR CONTROLS	→ SEE S335
	R235	IMPLEMENT CONTROLS	→
	R275	IMPLEMENT SAFETY	→ SEE S318
	R280	TRACTOR SAFETY	→
J943(2)	S276	SLOW MOVING VEHICLE EMBLEM	
J725	S277	LAMP BRACKET AND SOCKET	
J137(4)	S279	LIGHTING AND MARKING - FARM EQUIPMENT	
J955(7)	S297	FULLY SHIELDED POWER TAKE-OFF	
J389	S304	CONTROL SYMBOLS	→
J333(8)	R305	OPERATOR PROTECTION	→ SEE S383
J334(8)	R306	PROTECTIVE FRAME PERFORMANCE REQUIREMENTS	→ SEE S279
	S307	FLASHING WARNING LAMP FOR REMOTE MOUNTING	
J167	S310	PROTECTIVE FRAME - WITH OVERHEAD PROTECTION	
	R317	MOBILE SLURRY TANKS	
J208(2)	S318	SAFETY FOR AGRICULTURAL AND INDUSTRIAL WHEELED EQUIPMENT	
J841(1)	S335	OPERATOR CONTROLS ON AGRICULTURAL EQUIPMENT	→ SEE S383
J168(8)	S336	PROTECTIVE ENCLOSURES - TESTS AND PERFORMANCE	
	S338	SAFETY CHAIN FOR TOWED EQUIPMENT	
	EP342	SAFETY FOR ELECTRICALLY HEATED LIVESTOCK WATERERS	
	S347T	BLOWER PIPE AND CONNECTING FLANGE DIMENSIONS	
J284	S350	SAFETY ALERT SYMBOL FOR AGRICULTURAL EQUIPMENT	
	S354	SAFETY FOR FARMSTEAD EQUIPMENT	
	S355	SAFETY FOR AGRICULTURAL LOADERS	
	S361T	SAFETY FOR AGRICULTURAL AUGER CONVEYING EQUIPMENT	
	S362	WIRING FOR ELECTRICALLY DRIVEN OR CONTROLLED IRRIGATION MACHINES	
	EP364	INSTALLATION AND MAINTENANCE OF FARM STANDBY ELECTRIC POWER	
	S365T	BRAKE TEST PROCEDURES AND BRAKE PERFORMANCE ON AGRICULTURAL EQUIPMENT	
	EP367	FIELD SPRAYER CALIBRATION GUIDE	
	EP371	GRANULAR APPLICATOR CALIBRATION PROCEDURES	
	S373	SAFETY FOR SELF-UNLOADING FORAGE BOXES	
J1194	S383	ROPS FOR WHEELED AGRICULTURAL TRACTORS	
	EP381	LIGHTNING PROTECTION SPECIFICATIONS	
	S395	SAFETY FOR SELF-PROPELLED, HOSE DRAG AGRICULTURAL IRRIGATION SYSTEMS	

Year columns: '54 '55 '56 '57 '58 '59 1960 '61 '62 '63 '64 1965 '66 '67 '68 '69 1970 '71 '72 '73 '74 1975 '76 '77 '78-'79 '79-'80

(1) ALSO ANSI B114.7 (2) ANSI B114.3 (3) ANSI B114.1 (4) ANSI B114.6 (5) FIRST YEAR BOOK TO SHOW NUMBERS ON STANDARDS (6) DROPPED IN 1977 (7) DROPPED IN 1975 (8) DROPPED IN 1978

CARLTON L. ZINK 1980

800943

The Development of Agricultural Equipment Power Take-Off Mechanism

T. H. Morrell
Management Consultant
Owatonna, MN

THIS PAPER REFERS to present SAE Standards and Recommended Practices, and it reviews some early power take-off applications, developments leading to our present standards, some more evaluations to be considered and a summary.

Standards are very important to the agricultural industry to provide safety, interchangeability, and reliability at lower cost to the user. Before power take-off standards were attempted, the optimum drive line speed of tractors and power driven implements varied widely. The vertical and horizontal distances between the tractor drawbar hitch point and the tractor output shaft varied widely among all makes. The size of the output shaft varied and drive line dimensions on implements with respect to the drawbar hitch point also varied considerably.

Because of the lack of standards, the farm equipment industry had to provide an estimated 2500 special hook-up packages to match all makes of tractors with all makes of power driven implements. Customer complications can well be imagined:

(a) in selecting proper hook-up for tractor and implement combinations,

(b) in the added cost of many packages to adapt several different makes of implements to one or more makes of tractors, and

(c) in keeping the many packages properly matched to the various tractors and implements.

During the past 30 years, most tractors and implements have been manufactured according to the standards. There is no need to provide special hook-up packages to match these tractors and implements which are according to the standards.

The present SAE Power Take-Off

ABSTRACT

A power take-off mechanism was developed for farm equipment. Its purpose was to transmit power from the tractor engine to the propelled implement. The propelled implement could then perform its functions by utilizing the tractor engine power.

The power take-off (PTO) development has involved a large variety of tractors and implements and has included recommendations and standards for safety, interchangeability, and reliability. This paper outlines some of the important steps in the power take-off evolution from the first known application to present day standards.

The power take-off has provided great versatility in the mechanization of the farm, and this has helped to lower the overall cost of food production.

SAE/SP-80/470/$02.50

Standards and Recommended Practices for farm equipment are:

1. J1170 - SAE Standard (ASAE S203.9) "Rear Power Take-Off for Agricultural Tractors." (J1170 includes both 540 and 1000 RPM PTO standards, formerly J718 and J719). See Appendix "A".

2. J717 - SAE Recommended Practice (ASAE S333.1) "Auxiliary Power Take-Off Drives for Agricultural Tractors." See Appendix "B".

3. J721 - SAE Recommended Practice (ASAE S207.9) "Operating Requirements for Power Take-Off Drives." See Appendix "C".

4. J722 - SAE Recommended Practice (ASAE S205.2) "Power Take-Off Definitions and Terminology for Agricultural Tractors." See Appendix "D".

The American Society of Agricultural Engineers (ASAE) publishes the above standards and recommended practices in its Yearbook. ASAE also publishes some standards which affect only the implements. Implements are not considered within the scope of the Society of Automotive Engineers standards. These additional ASAE standards are:

1. ASAE S331.2 "Implement Power Take-Off Drive Line Specifications." This standard establishes 6 categories of universal joint drive lines with respect to static and dynamic torsional requirements. These 6 categories include a wide range of implements from low to high power requirements. The drive line to the tractor output shaft is the responsibility of the implement manufacturer.

2. ASAE S314 "Implement Power Take-Off and Drive Line Pedestal Shafts." This standard provides specifications for 1 3/8" and 1 3/4" PTO shafts for uniform driving means, retaining means and drive line positioning.

Since the late 1930's, most of the power take-off standards and recommended practices have been proposed by the Farm and Industrial Equipment Institute (Formerly Farm Equipment Institute) Engineering Committees. When these proposals are acceptable and approved, those affecting tractors and implements are published by ASAE in its Yearbook and those affecting tractors are published by SAE in its Handbook.

(Fig. 1) shows a typical power take-off mechanism of the late 1970's which is according to SAE J1170.

Now, let us go back to some early power take-off applications and then follow the important developments to today's PTO's.

Fig. 1 - Typical application of J1170 PTO standard

EARLY POWER TAKE-OFF APPLICATIONS

(Fig. 2) shows one of the earliest known tractor rear PTO applications in the United States. It was a Jeep-like machine shipped by Renard from England in 1904 around Cape Horn to the West Coast of the United States. It had a power take-off shaft from the tractor to a driving axle on each of three trailers. The ship encountered a storm, and one of the trailers on the ship's top deck was lost at sea. Along with the "20 Mule Team", the tractor and two trailers were used for many years to haul borax from Death Valley.

The beginning of the use of power take-off and the Society of Automotive Engineers was almost simultaneous. Also, we can declare a 75th Annivers-

Fig. 2 - Renard tractor - 1904

ary for the power take-off.

In 1918, International Harvester introduced its 8-16 Kerosene tractor which had a power take-off as optional equipment. This was the first known American tractor to offer a practical rear PTO mechanism for propelling trailed implements.

(Fig. 3) shows the IHC 8-16 tractor.

In 1921, International Harvester was the first to provide the rear PTO as standard equipment on its new McCormick-Deering 15-30 tractor.

(Fig. 4) shows the IHC 15-30 tractor.

The PTO became popular in approximately 1923.

In 1928, Hart-Parr advertised the first independent power take-off on its Model 18-36. The 18-36 was tested at Nebraska, October 1926, Report #128.

(Fig. 5) shows the Hart-Parr 18-36. Hart-Parr became a part of a new company, Oliver Farm Equipment Company, in 1929. In 1970, Oliver became a part of White Farm Equipment Company, Division of White Motor Corporation.

Fig. 3 - International Harvester 8-16 Kerosene tractor

Fig. 4 - McCormick-Deering 15-30 tractor

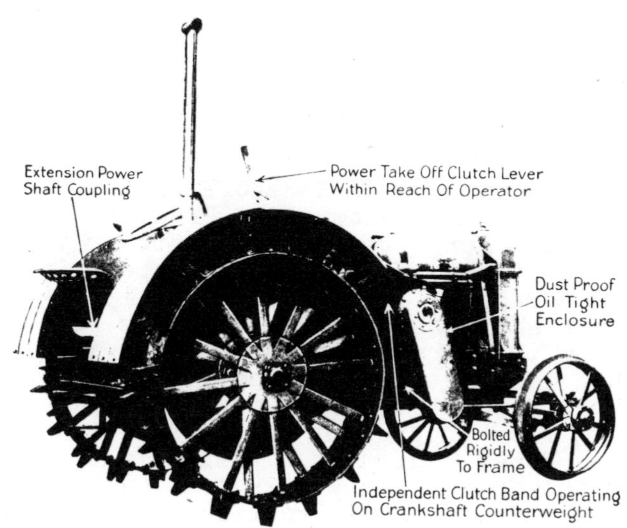

Fig. 5 - Hart-Parr 18-36 tractor

EARLY SAE PTO STANDARDS AND RECOMMENDED PRACTICES

1918 - Tractor drawbar standard height.
"Vol. 1 S. A. E. Data Sheet 55
HEIGHT OF TRACTOR DRAWBAR
S. A. E. Standard

The standard height of drawbar shall be 17 in. for both plowing and other work."

The drawbar is to become related to PTO standard at a later date.

1924 - June - A report from SAE Agricultural Power Equipment Division recommended 536 RPM as standard speed and with clockwise rotation when viewed from the rear.

"Vol. XIV June, 1924 No. 6
TRACTOR POWER TAKE-OFF SPEED

At the meeting of the Agricultural Power Equipment Division in April, the following recommendation was approved:

> The normal speed of the power take-off of tractors designed for operating tractor-propelled agricultural implements shall be 536 r.p.m., the rotation to be clockwise when looking in the direction in which the tractor travels.

This recommendation was approved in view of the fact that the power take-off speed used throughout the industry has been practically constant for 30 years, this being determined by the speed of the sickles. It is believed to be advisable, however, to recognize this standard speed for the future guidance of tractor and implement designers. The Division therefore submits the proposal for adoption as S. A. E. Recommended Practice."

1924 - July - The report was approved for adoption as an SAE Recommended Practice with the addition of the tolerance of ± 20 RPM.

1924 - August - Tractor Power Take-Off Speed Recommended Practice was published.

"Vol. 1 S. A. E. Handbook E53

TRACTOR POWER TAKE-OFF SPEED

S. A. E. Recommended Practice

The normal speed of the power take-off of tractors designed for operating tractor-propelled agricultural implements shall be 536 plus or minus 20 r.p.m., the rotation to be clockwise when looking in the direction in which the tractor travels.

From the report of the Agricultural Power Equipment Division, adopted by the Society July 1924."

1929 - Recommended Practice was revised to include 6B spline, 3" long for 1 1/8" and 1 3/8" shafts.

"176

TRACTOR POWER TAKE-OFF SPEED

S. A. E. Recommended Practice

The power-take-off shaft on the tractor shall be provided with an S.A.E. 6B spline fitting. The straight length at the root of the spline shall be 3 in. Retaining means for securing the fitting shall not project more than 1 in. from the end of the spline.

The normal speed of the power-take-off shaft shall be 536 r.p.m., plus or minus 20, the direction of rotation to be clockwise when facing in the direction the tractor travels.

Two sizes of power-take-off shaft shall be used, as follows:

(1) The 1 1/8-in. splined shaft on tractors with engines developing up to 20 b-hp.

(2) The 1 3/8-in. splined shaft on tractors with engines developing up to 40 b-hp.

These ratings are based on the use of material having a minimum torsional yield-point of 65,000 lb. per sq. in.

From the report of the Agricultural Power Equipment Division, adopted by the Society July 1924. Revised February 1929."

1938 - The PTO Recommended Practice became a standard in January 1938. Also, it was revised to include shaft dimensions, 1 3/4" splined shaft, 3¼" spherical clearance radius, drawbar relationship to output shaft, shielding of drive line and a reference to adaptor parts to hitch implements to standard tractor PTO.

In 1937, the American Society of Agricultural Engineers (ASAE) adopted the PTO Standard but included another shaft size of 1 3/4" diameter and additional data as to the location of PTO shaft with respect to drawbar and other requirements. Since this time, ASAE and SAE have coordinated and published basically the same PTO standards. The complete 1938 Standard is Appendix "E".

This was the first SAE report in the PTO development concerning shielding of the drive line. "The tractor manufacturer shall adequately shield the power take-off shaft and tractor universal joint, and provide protection for the operator against the telescoping member attached thereto, assuming connection between tractor and implement is according to recommended practice."

"The manufacturer of a power take-off driven machine shall furnish the power drive parts up to the tractor spline shaft, the necessary hitch parts to attach to the recommended drawbar location, and all shields, except the one attached to the tractor, covering the spline shaft fitting or universal joint."

"The tractor power take-off drive shall be provided with a throw-out clutch, operating independently of the tractor travel, of a design safe against accidental engagement and with a control located conveniently to the operator."

During the 1920's PTO drive lines were generally unprotected.

(Fig. 6) shows a typical PTO drive line in the 1920's.

During the early 1930's, some inverted "U" tunnel type telescoping

Fig. 6 - PTO shaft on grain binder of 1920's

shields were adopted.

(Fig. 7) shows the tunnel type shields and how they were attached with a latch to a tractor master shield and the implement shield. The tractor master shield was included in the 1946 SAE Handbook printing.

Fig. 7 - Inverted "U" tunnel type shield

IN 1946 - (Fig. 8) shows the revised standard to include views and dimensions of PTO master shield.

(Fig. 9) shows the agricultural tractor drawbar standard which established more precise drawbar and PTO output shaft relationship.

(Fig. 10) shows a drawing of PTO shaft and hub or coupling. Drawings

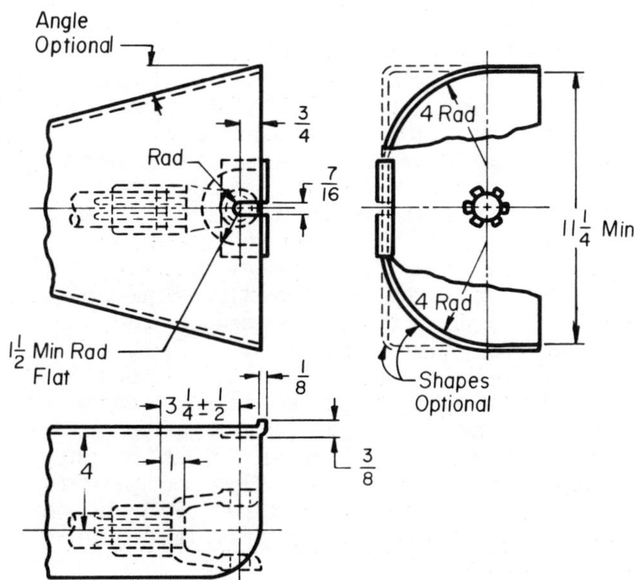

Fig. 8 - Tractor PTO master shield

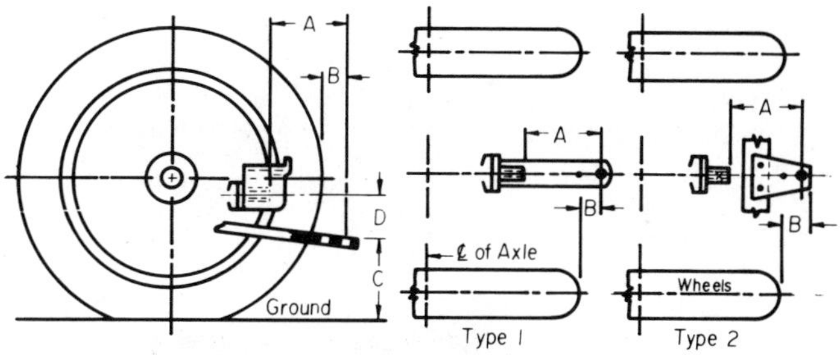

Fig. 9 - Tractor PTO output shaft and drawbar relationship

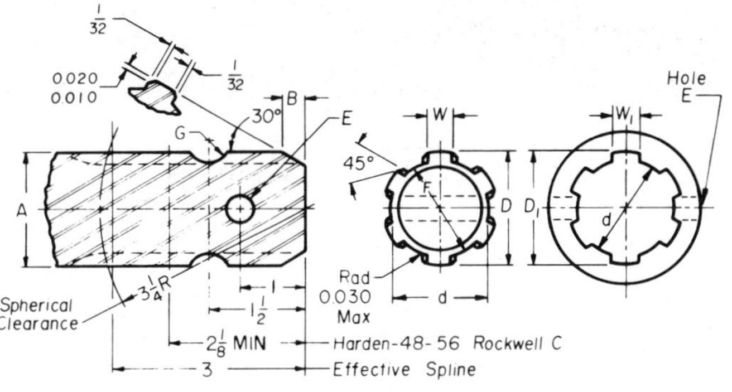

Fig. 10 - Outline drawing of 1 3/8" 6B splined shaft and coupling

of the power take-off and drawbar relationship were included in the drawbar standard. 1 1/8" shaft standard was discontinued. See appendices "F" and "G".

ENGINE POWERED DRAWN EQUIPMENT

By 1946, pull type implements, such as combines and hay balers, were driven by a separate engine on the implement and were pulled behind a tractor. An example is shown in (Fig. 11).

This photo of Allis Chalmers tractor and New Holland baler was taken in approximately 1948.

The reason for the separate engine was to permit stopping and starting of the forward travel without interrupting the power to the implement. When a sudden overload of material to the implement was encountered, it was necessary to stop forward motion so that the overload could be cleared without clogging the implement.

The extra cost and complications of adding an engine to the implement prompted the re-introduction of the independent PTO.

INDEPENDENT POWER TAKE-OFF

The independent PTO provided the

Fig. 11 - Baler with engine to drive its baling mechanism

Fig. 12 - Oliver's tractor independent PTO

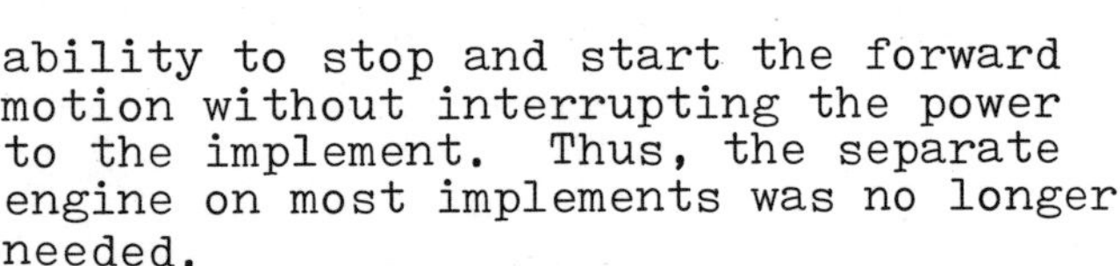

ability to stop and start the forward motion without interrupting the power to the implement. Thus, the separate engine on most implements was no longer needed.

In approximately 1946, Oliver Corporation and Cockshutt provided their new lines of tractors with independent PTO. Other tractor manufacturers followed suit soon afterwards.

(Fig. 12) shows the general configuration of Oliver's independent PTO.

(Fig. 13) shows a cross section of the Oliver independent PTO and its drive.

FURTHER 540 RPM POWER TAKE-OFF DEVELOPMENTS

In approximately 1950, the standard was changed from 536 ± 10 RPM to 540 ± 10 RPM.

The 540 RPM PTO standard has continued much the same since 1948. There have been some dimensions added, some editorial changes, and some improvements in safety. Some of these developments will be reviewed later in this paper.

In 1953, SAE J721 - SAE Recommended Practice "Operating Requirements for Power Take-Off Drives" was approved. The purpose of this Recommended Practice was to provide guidelines for the design of drive lines between the tractor and the implement. It includes:

a. Instructions to the operator.

b. Implement hitch and power line design requirements.

c. Maximum bending load limitations for power take-off drives employing v-belts and chains.

d. Power line protective couplings and maximum torsional load limitations for power take-off shafts.

During the late 1940's and early 1950's, many new PTO driven implements, such as mounted corn pickers, became quite common. The drive was often by a sprocket and chain or a belt from the tractor PTO shaft. These applications and the independent PTO caused some new problems. The radial type loading on the tractor PTO shaft was not anticipated early in the development of the PTO standards.

Transmission driven PTO's were normally engaged smoothly through the tractor main clutch. The independent PTO could be engaged quickly without an apparent "jolt" to the operator. Such quick engagement could cause extreme shock loads on PTO mechanism. Radial loaded shafts were often bent from these shock loads. Also, the independent PTO "opened the door" to introduction of high power receiving implements, such as forage harvesters.

The variety of tractor shaft to drawbar vertical relationship and adjustment of the implement pedestal bearing height permitted very poor drive line alignment in field operations. The poor alignment often resulted in early failures of universal joints and other drive line components.

These problems and new PTO driven implement applications prompted the development of J721.

This Recommended Practice has had many editorial changes to keep it up-to-date. One significant change was

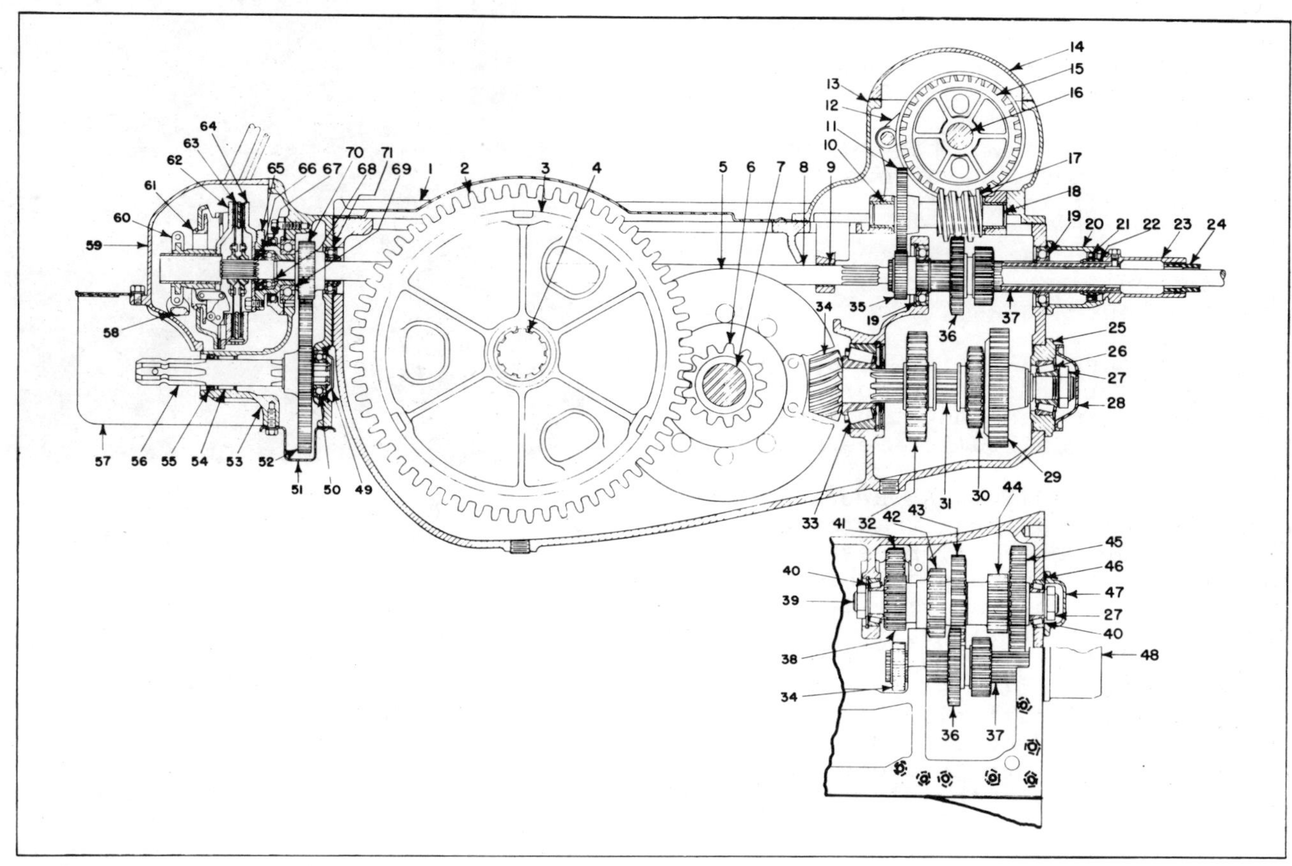

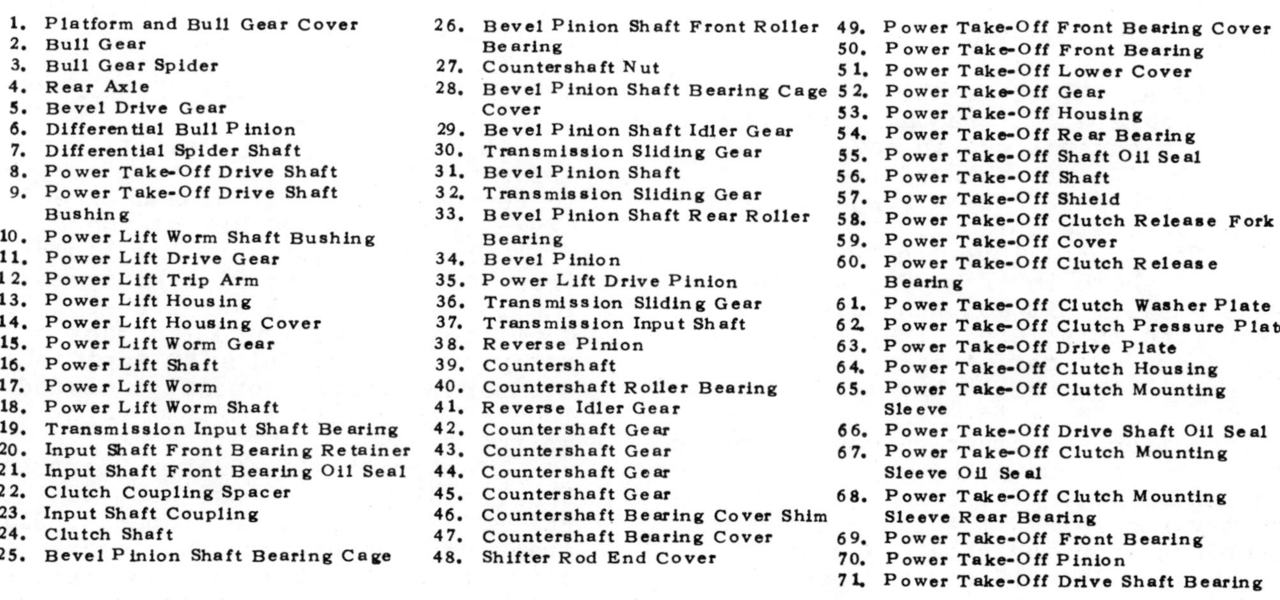

1. Platform and Bull Gear Cover
2. Bull Gear
3. Bull Gear Spider
4. Rear Axle
5. Bevel Drive Gear
6. Differential Bull Pinion
7. Differential Spider Shaft
8. Power Take-Off Drive Shaft
9. Power Take-Off Drive Shaft Bushing
10. Power Lift Worm Shaft Bushing
11. Power Lift Drive Gear
12. Power Lift Trip Arm
13. Power Lift Housing
14. Power Lift Housing Cover
15. Power Lift Worm Gear
16. Power Lift Shaft
17. Power Lift Worm
18. Power Lift Worm Shaft
19. Transmission Input Shaft Bearing
20. Input Shaft Front Bearing Retainer
21. Input Shaft Front Bearing Oil Seal
22. Clutch Coupling Spacer
23. Input Shaft Coupling
24. Clutch Shaft
25. Bevel Pinion Shaft Bearing Cage
26. Bevel Pinion Shaft Front Roller Bearing
27. Countershaft Nut
28. Bevel Pinion Shaft Bearing Cage Cover
29. Bevel Pinion Shaft Idler Gear
30. Transmission Sliding Gear
31. Bevel Pinion Shaft
32. Transmission Sliding Gear
33. Bevel Pinion Shaft Rear Roller Bearing
34. Bevel Pinion
35. Power Lift Drive Pinion
36. Transmission Sliding Gear
37. Transmission Input Shaft
38. Reverse Pinion
39. Countershaft
40. Countershaft Roller Bearing
41. Reverse Idler Gear
42. Countershaft Gear
43. Countershaft Gear
44. Countershaft Gear
45. Countershaft Gear
46. Countershaft Bearing Cover Shim
47. Countershaft Bearing Cover
48. Shifter Rod End Cover
49. Power Take-Off Front Bearing Cover
50. Power Take-Off Front Bearing
51. Power Take-Off Lower Cover
52. Power Take-Off Gear
53. Power Take-Off Housing
54. Power Take-Off Rear Bearing
55. Power Take-Off Shaft Oil Seal
56. Power Take-Off Shaft
57. Power Take-Off Shield
58. Power Take-Off Clutch Release Fork
59. Power Take-Off Cover
60. Power Take-Off Clutch Release Bearing
61. Power Take-Off Clutch Washer Plate
62. Power Take-Off Clutch Pressure Plate
63. Power Take-Off Drive Plate
64. Power Take-Off Clutch Housing
65. Power Take-Off Clutch Mounting Sleeve
66. Power Take-Off Drive Shaft Oil Seal
67. Power Take-Off Clutch Mounting Sleeve Oil Seal
68. Power Take-Off Clutch Mounting Sleeve Rear Bearing
69. Power Take-Off Front Bearing
70. Power Take-Off Pinion
71. Power Take-Off Drive Shaft Bearing

Fig. 13 - Cross section of Oliver's tractor transmission, differential, final drive and independent PTO

the addition of a table showing vertical drawbar load limitations with various size classes of tractors. See Appendix "C".

In 1954, SAE J722 - Recommended Practice "Power Take-Off Definitions and Terminology for Agricultural Tractors" was approved. This Recommended Practice defines the various types of power take-off including:

a. Clutch, master
b. Power take-off, tranmsission driven
c. Power take-off, continuous running
d. Power take-off, independent

It establishes a uniform terminology within the industry for the various types of PTO's. See Appendix "D".

In 1955, J717 - SAE Recommended Practice "Agricultural Tractor Auxiliary Power Take-Off Drives" was approved. The purpose of this Recommended Practice was to establish speed, shaft dimensions, and direction of rotation for mid and side PTO's. There was some interest in auxiliary mid, right and left side, tractor power outlets. This Recommended Practice provided guidelines to avoid future problems, such as non-interchangeability.

This Recommended Practice has remained much the same except for editorial changes to keep it up-to-date and to add location information and shielding recommendations. See Appendix "B".

In 1958, the integral rotating power take-off drive line shield was proposed.

(Fig. 14) shows the integral rotating shield. It was accepted and put into general use in early 1960's. The 1 3/4" shaft changed from 1 3/4" 6 B spline to 27 teeth 16/32 diametral pitch involute spline.

The tunnel inverted "U" type shield, shown in (Fig. 7), was not a good answer. It got bent easily if a rear tire interfered with it on turns. Any dents or rust caused problems to telescope the shield and to connect it to the tractor master shield. Such malfunction and other reasons prompted the user to remove the shield and never re-install it. The farm equipment industry recognized the hazards and the problems when the shield was omitted. During the late 1940's and early 1950's, some companies started to provide an assembly whereby the shield was journaled on the telescoping drive line shaft. The telescoping shield had a smooth outer surface. The shield could be held while the shaft inside

Fig. 14 - Integral rotating shield on PTO telescoping drive line

was turning. It had the advantage that it was a part of the PTO telescoping drive shaft assembly and was not easily removed. This afforded much better long term protection for the user.

Other means of shielding the drive line have been considered, such as a fully shielded assembly, which was used for a period of time by some and has been discontinued for reliability reasons. This was SAE J955 Recommended Practice "Full Shielding of Power Drive Lines for Agricultural Implements and Tractors". SAE J955 (ASAE S297T) was cancelled in approximately 1974. See Appendix "H".

1000 RPM POWER TAKE-OFF DEVELOPMENT

By the early 1950's, it became apparent that tractors would continue to become more powerful and exceed the capacity of the 540 RPM standard.

The Farm Equipment Institute (now Farm and Industrial Equipment Institute - FIEI) Power Take-Off Sub-Committee was assigned the task of studying and proposing an additional standard. Their proposal included 1000 RPM PTO with 1 3/8" involute spline.

In April 1957, Oliver Corporation, Charles City, Iowa, was host to the FEI PTO Sub-Committee. Special Oliver tractors were prepared to check-out drive lines for tractor and implement relationships of proposed 1000 RPM power take-off.

(Fig. 15) shows the special tractor adaptation.

The front wheels of an Oliver row-crop tractor were removed. In place

Fig. 15 - Special tractor combination for studying PTO's

of the front wheels, a special adaptor was attached to tractor front frame and to the forks of a fork lift tractor. The purpose of this special arrangement was to provide a means of simulating a tractor and PTO driven implement negotiating ditches and rough terrain and while turning. With respect to the level ground, the tractor front end could be lowered approximately 17° before the front adaptor touched the ground. The tractor front end could be elevated approximately 20 to 24° before the standard 15" high drawbar and implement hitch touched the ground. A protractor mounted on the side of the tractor provided quick readings of the tractor's vertical incline.

(Fig. 16) shows how the tractor was angled with respect to the implement. When in the inclined and turned positions, they simulated extreme operating conditions. This set-up permitted much to be learned with regard to drive shaft telescoping capabilities and universal joint applications.

Fig. 16 - Study of PTO drive line while turning and tractor front end elevated

These tests indicated that the 540 RPM standard of 14" horizontal distance between drawbar hitch point and tractor output shaft and 14" horizontal distance from hitch point to implement power shaft did not permit operation over extreme rough terrain. The telescoping shaft could bottom on a short coupled tractor and implement combination when the front end of the tractor was inclined upward approximately 15° and during a sharp turn. The telescoping shaft could almost separate on a short coupled tractor and implement combination when tractor was straight ahead and tractor front end inclined downward approximately 17°.

The 1000 RPM PTO drive line at 14" horizontal distance from tractor PTO output shaft to hitch point and during sharp turns vibrated more than the 540 RPM PTO drive line. Further studies were conducted using 16" and 18" horizontal distances. A 16" horizontal dimension from the tractor output shaft to the drawbar hitch point provided much better telescoping under these extreme conditions. The 16" horizontal dimension was established for the 1 3/8" involute splined shaft at 1000 RPM.

Tractor front end stability, clearance with rear tires during turns, weight of implement tongues on tractor drawbar and many others were considerations to limit how far the drawbar hitch point could be located rearward of the tractor.

Another interesting situation occurred during this special set-up. When a standard drawbar was hitched to a standard implement clevis and a recommended hitch pin size was used, the pin would bend when the tractor front end was inclined approximately 5°. In order to provide the flexibility for the tractor and implement evaluation, a special drawbar with 3 point hitch ball was prepared and used for the demonstrations.

In August 1957, a field demonstration was planned near Morton, Illinois, to test and demonstrate the 1000 RPM PTO units from various companies.

(Fig. 17) shows a typical field demonstration of 1000 RPM PTO.

The demonstration consisted of tractors from most companies and implements which were representative of various types from small and short coupled as shown in (Fig. 18) to long coupled units as shown in (Fig. 17).

Fig. 17 - 1000 RPM PTO field evaluation

Fig. 18 - Typical short coupled tractor and PTO driven implement

These tractors and implements were equipped with the company's proposed 1000 RPM PTO. The designs were generally refined, based upon what was learned in April 1957, at Charles City, Iowa. The designs were further refined based upon the results of the field review and tests at Morton, Illinois, in August 1957. Additional 1000 RPM PTO standards review meetings were conducted during 1957.

In April 1958, the SAE Tractor Technical Committee approved the 1000 RPM PTO Standard J719 - with 1 3/8" dia. 21 teeth, 16/32 diametral pitch involute splined shaft.

The industry wide participation in the 1000 RPM PTO evaluations revealed many other considerations to provide an effective future Power Take-Off program.

OTHER CONSIDERATIONS

EXCESSIVE PTO SPEEDS - During the 1950's, tractor engine speeds were increased. The overall configuration of some tractors did not permit a greater reduction to the PTO output shaft. The engine governed speed for drawbar work was higher than the engine speed to drive the PTO implement.

A survey was conducted and it was learned that some early implements would be subject to failure and safety hazards if over speeded. In June 1959, this subject was reviewed and later an editorial change was approved. J1170 paragraphs 2.4 and 2.5 limit the over speeding of 540 and 1000 RPM PTO driven implements. See Appendix "A".

The non interchangeable spline and adequate decals were means of greatly reducing the possibility of propelling a 540 RPM implement at 1000 RPM.

CONVERSION OF TRACTORS AND IMPLEMENTS FROM 540 to 1000 RPM PTO'S - Each company was responsible for providing information to the field on tractors and implements that could be converted from 540 to 1000 and 1000 to 540 RPM PTO.

LOW FRICTION TELESCOPING DRIVE LINE - High power receiving implements, such as roto tillers and field silage harvesters, experienced early drive line failures caused by high end thrust when operated at continuous full power over rough terrain and during sharp turns. Some experimental ball recirculating telescoping drive lines which were tested in 1958 solved the problem. This experience led to the recommendation to provide low friction telescoping drive lines for implements capable of utilizing continuous torque in excess of 3000 lb. in. (34.6 kg-m). The low friction telescoping drive shaft is to be incapable of transmitting an axial thrust in excess of 1500 lbs. (680 kg). See SAE J721 Appendix "C".

Some tests indicated that at 50 HP, statically, it took 19,000 lbs. end thrust to telescope a conventional square shaft assembly. A ball recirculating shaft assembly only required 420 lbs. Also, the ball recirculating telescoping drive shaft was quieter and had much less vibration during sharp turns.

IMPROVED DRAWBAR HITCH POINT FASTENING - A simple drawbar ball hitch and some proposals for fast hitching were reviewed in 1958. The intent of these proposals was to provide more flexibility to avoid binding and bending of the hitch pin. None of these appeared too satisfactory in view of the millions of PTO implements, in customer's hands, which would cause hitching problems if these proposals were adopted.

IMPROVED FASTENING OF FORWARD UNIVERSAL JOINT TO TRACTOR PTO SHAFT – Considerable early failures of retaining pins on universal joints were reported in 1958. Most of these were caused by excessive end thrust on the drive line from operating over rough terrain and under load on sharp turns. Several experimental attempts were made to solve this problem through a different attaching means. None of these appeared to solve the problem.

PTO TELESCOPING DRIVE SHAFT ASSEMBLY STANDARD - An attempt was made, starting in 1959, to provide a standardized ball recirculating telescoping shaft assembly. The objectives were to solve many types of drive line failures and to provide an economical assembly. One of the objections to the ball recirculating type was its high cost. If a standard telescoping drive line could be provided, then instead of having a telescoping shaft assembly on each implement, one low friction telescoping shaft could be provided for several implements for any one tractor.

This proposal was dropped because of interchangeability problems with implement drive lines on millions of implements already in the hands of customers.

IMPROVED SHIELDING OF FORWARD UNIVERSAL JOINT - Since the 1920's, there has been a continuous effort to improve the drive line shielding. In 1959, many ideas were proposed. Some experimental ones were made and tested. These considerations resulted in the Recommended Practice SAE J955 "Full Shielding of Power Drive Lines for Agricultural Implements and Tractors" approved in 1966 and was discontinued in approximately 1974.

GROUND SPEED PTO FOR IMPLEMENTS AND TRACTORS - There were considerable discussions about providing a ground speed PTO for rakes, seeders, planters, fertilizers, etc. This requirement was primarily for European countries and was being considered for an International Standards Organization (ISO) standard. The proposal was for 19½" ± 1" for each ground speed PTO revolution.

This subject was tabled for lack of interest by United States industry and customers.

OTHERS - In 1967, 1 3/4" dia. 20 teeth, 12/24 diametral pitch was approved for 1000 RPM PTO.

In 1978, the 540 RPM SAE Standard J718 and the 1000 RPM SAE Standard J719 were combined into the new J1170 Standard.

J1170 included the addition of a larger tractor master shield for the 1 3/4" 1000 RPM PTO shaft. A larger shield is necessary for clearance with the larger universal joints needed for drive lines transmitting power in excess of 100 HP.

Also, J1170 refers to J208 SAE Standard (ASAE S318) "Safety for Agricultural Equipment". The reference is in regard to safety when tractor master shield is omitted for any reason. See Appendix "I".

In 1980, there is now some concern about unequal angles of universal joints during sharp turns. Some designs for 1 3/8" 1000 RPM drive line have a horizontal dimension from tractor PTO shaft to drawbar hitch point of 16" and from drawbar hitch point to implement shaft of 18" or more. These unequal angles cause excessive vibration during sharp turns and particularly if there is end thrust on the drive line.

SOME INDUSTRY EVALUATIONS TO BE CONSIDERED

GROWTH OF TRACTOR SIZE AND POWER - Tractors and implements have grown in size at a rapid pace since the late 1950's. To keep up with effective standards has been a major task and has required continuous committee efforts. These efforts must continue as long as there is growth.

The 540 RPM standard is capable of transmitting an approximate maximum of 50 PTO horsepower continuously.

The 1000 RPM 1 3/8 involute shaft standard is capable of transmitting an approximate maximum of 100 PTO horsepower continuously.

The 1000 RPM 1 3/4 involute shaft standard is capable of transmitting an approximate maximum of 200 PTO horsepower continuously.

MATCHING OF TRACTORS AND IMPLEMENTS - The matching of large tractors to small power requirement implements, such as a mower equipped with 540 RPM capability, can be a serious problem. The other extreme, such as small tractors to forage harvesters, can also be a problem.

Careful instructions and non-interchangeable drives assist in proper match of PTO speeds. There was an estimated quantity of 4 million 540 RPM

PTO driven implements in the hands of U. S. customers when the 1000 RPM PTO standard was approved.

END THRUST ON DRIVE LINE DURING TURNS UNDER FULL POWER - Excessive end thrust on the drive line can occur on some tractor and implement combinations. An example is a field silage harvester which can absorb full power of the tractor while turning. Excessive end thrusts and early drive line component failures can result if such a combination is not equipped with a low friction telescoping drive line. The conventional square shaft and square tube tends to lock up and not telescope when submitted to continuous high torque load. A very high percent of field failures of PTO universal joints, drive lines, etc. are caused by excessive end thrust.

DRAWBAR HITCHES FOR TRACTOR AND IMPLEMENTS - Many different designs have been considered which would improve the drawbar hitches. The present standard lacks flexibility for negotiating rough terrain with short coupled implements. The hitch pin tends to bind and bend. Also, one man hitching is often difficult to accomplish.

SUMMARY

During the last 75 years, the application, safety, versatility, capability and quality of power take-off drives have improved continuously. The PTO has resulted in one of the greatest practical innovations on farm equipment.

The committees who have worked diligently to keep up with the evolution of farm equipment PTO applications are to be complimented. The history of PTO standards and recommended practices indicate the efforts required to effect safety, interchangeability and reliability in a field which is rather complex. These efforts will need to continue as long as there is growth in power of tractors and size of implements and wider scope of applications. There is still much to be considered.

No doubt, the cost of food production is greatly reduced through the use of the versatile PTO Drives.

ACKNOWLEDGEMENTS

The writer wishes to express appreciation for the information he has received from Mr. T. E. Martin and Mr. M. D. Stewart (both retired from Oliver Corporation); Mr. E. W. Tanquary (retired from International Harvester Company) who provided some PTO background considerations; and Mr. Robert C. Uhl (Staff Engineer for SAE) who provided some historical SAE records. He further wishes to acknowledge those who provided additional information and photos: Mr. G. Lennes (International Harvester Company), photo of IHC 8-16 and 15-30; Mr. M. J. Happe (New Holland), photo of Allis Chalmers tractor and New Holland baler; Mr. P. E. Lockie and Mr. M. W. Thelen (White Farm Equipment), photos and historical information on Hart-Parr and Oliver tractors; and Mr. W. H. Worthington (Wayne Engineering of Cedar Falls, Iowa), Renard photo. He also wishes to give credit to his wife Blanche for her able assistance in the preparation of this paper.

REFERENCES

1. R. B. Gray, "Development of the Agricultural Tractor in the United States."

2. Pierce Fulkerson, "Understanding the Modern Farm Tractor."

3. "A History of Man's Progress-From 1830 to the Present." The Harold Warp Pioneer Village, Minden, Nebraska, 1978.

4. John T. Schlebecker, "Whereby We Thrive - A History of American Farming, 1607 to 1972."

5. SAE, ASAE and FIEI Papers

   a. ASAE Paper, L. A. Gilmer, "Direct Drive or Independent Power Take-Off", May 1948.

   b. ASAE Paper, Merlin Hansen, "Loads Imposed on Power Take-Off Shafts by Farm Implements", December 1951.

   c. ASAE Report, T. H. Morrell, "A Study of Power Take-Off Drives", December 1951.

   d. ASAE Paper, D. E. Burrough, "Power and Torque Distribution in Farm Machine Drive Shafts", June 1953.

   e. SAE Paper, F. M. Potgieter, "Universal Joint Applications to Power Take-Off Drives on Farm Machinery", September 1953.

   f. FIEI PTO Committee, J. H. Bornzin, "A New Look at PTO Drive Lines", February 1963.

   g. ASAE and FIEI Paper, T. H. Morrell, "Industry Standardization of Farm Equipment", December 1965.

   h. ASAE Paper, J. H. Bornzin, "Tractor to Implement-Power Drive Shafts", December 1970.

   i. ASAE Paper, J. H. Bornzin, "Toward a Safer PTO", Agricultural Engineering Journal, July 1973.

6. Farm and Industrial Equipment Institute Meeting Minutes.

7. SAE Handbooks.

APPENDIX A

REAR POWER TAKE-OFF FOR AGRICULTURAL TRACTORS—SAE J1170

SAE Standard

Report of Tractor Technical Committee approved December 1976. Conforms to a corresponding American Society of Agricultural Engineers Standard. Conforms to a report of the Farm and Industrial Equipment Institute. (This document incorporates information formerly contained in SAE J718d and SAE J719d.) Editorial change May 1977.

1. Purpose and Scope

1.1 This standard establishes the specifications and dimensions that are essential in order that 540 r/min and 1000 r/min power take-off-driven machines may be operated with any make of tractor equipped with the corresponding power take-off drive.

1.2 This standard does not, in itself, insure adequate telescoping of the power line or safety shielding.

2. Specifications

2.1 The power take-off shaft (except belt-pulley shafts) for power take-off drives extending from the tractor to the rear shall have the dimensions shown in Figs. 1, 2, 3 and 4 and Table 1.

2.2 Dimensions associated with the drawbar and operating dimensions shall conform to Fig. 5 and Table 2.

The circumferential groove is provided for a locking means in the implement hub.

Effective spline length, B, to be heat treated for surface durability (within Rockwell C 48-56).

FIG. 1—POWER TAKE-OFF SHAFT (SEE TABLE 1)

2.3 The two rated operating speed classifications of the power take-off shaft, when under normal load, shall be 540 ± 10 r/min and 1000 ± 25 r/min. The direction of the rotation shall be clockwise when facing in the direction of forward travel.

2.4 A means to indicate when the power take-off is operating at normal speed shall be provided on tractors capable of driving the 540 r/min shaft in excess of 600 r/min and the 1000 r/min shaft in excess of 1100 r/min.

2.5 Tractors capable of driving the 540 r/min shaft in excess of 630 r/min and the 1000 r/min shaft in excess of 1170 r/min shall also include a suitable warning of operation in excess of those speeds.

2.6 Power drive line shielding and tractor master shield shall conform to SAE Standard J208—Safety for Agricultural Equipment. Tractor master shield dimensions shall conform to Fig. 6.

2.7 If removal of the tractor master shield is required for integral PTO driven implements, shielding shall be provided with the implement to provide protection as specified in SAE J208.

2.8 The tractor power take-off shaft shall be covered at all times, either by a master shield or other protective means, when not connected to a drive assembly.

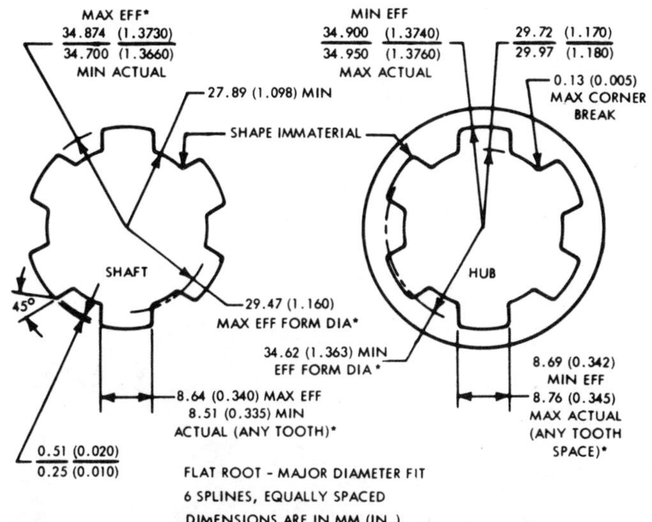

FIG. 2—540 R/MIN POWER TAKE-OFF (35 mm (1⅜ in) DIAMETER STRAIGHT SIDE SPLINE DIMENSIONS)

2.9 The drawbar hitch point shall be directly in line with the centerline of the tractor power take-off shaft, and provisions shall be made on the tractor for locking the drawbar in this position.

2.10 The location of the tractor power take-off shaft shall be within the limits of 25.4 mm (1 in) to the right or left of the centerline of the tractor, tractor centerline being the recommended location.

2.11 The tractor drawbar when swinging through a 1.57 rad (90 deg) turn right or left, shall clear an implement clevis with the following dimensions:

2.11.1 35 mm (1⅜ in) power shaft: 76 mm (3.0 in) vertical opening, and 76 mm (3.0 in) throat depth from hitch pin center.

2.11.2 45 mm (1¾ in) power shaft: 102 mm (4.0 in) vertical opening, and 102 mm (4.0 in) throat depth from hitch pin center.

TABLE 1—POWER TAKE-OFF SHAFT DIMENSIONS[a] (SEE FIG. 1)

		35 mm (1-3/8 in) Dia. 540 r/min	35 mm (1-3/8 in) Dia. 1000 r/min	45 mm (1-3/4 in) Dia. 1000 r/min
A	Groove to end of shaft	38.1 (1.50)	25.4 (1.00)	38.1 (1.50)
B	Effective spline length with relation gage, min	76.2 (3.00)	63.5 (2.50)	88.9 (3.50)
C	Chamfer	7.1 (0.28)	4.8 (0.19)	7.6 (0.30)
D	Chamfer angle	0.5 rad (30°)	0.5 rad (30°)	0.5 rad (30°)
E	ID of groove	29.46 (1.160) / 29.26 (1.152)	29.46 (1.160) / 29.26 (1.152)	37.34 (1.470) / 37.13 (1.462)
F	Radius of groove	6.86 ± 0.25 (0.270 ± 0.010)	6.86 ± 0.25 (0.270 ± 0.010)	8.38 ± 0.25 (0.330 ± 0.010)
G	Spherical clearance radius on tractor, min	82.6 (3.25)	82.6 (3.25)	101.6 (4.00)
H	Location of center of clearance radius	0	12.7 (0.50)	0
J	Break sharp corner or chamfer	Yes	Optional	Optional

[a]Dimensions are in mm (in) except where indicated otherwise.

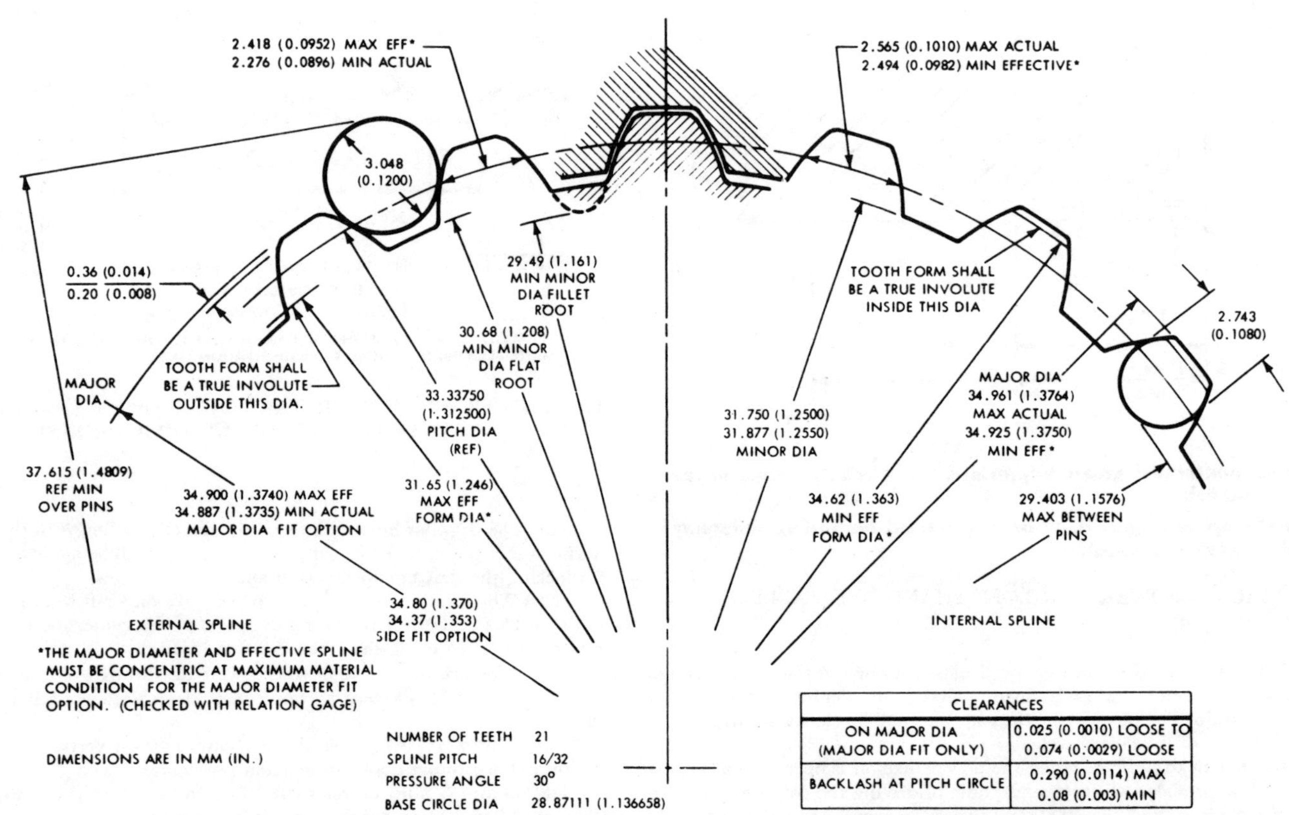

FIG. 3—1000 R/MIN POWER TAKE-OFF (35 mm (1⅜ in) INVOLUTE SPLINE DIMENSIONS)

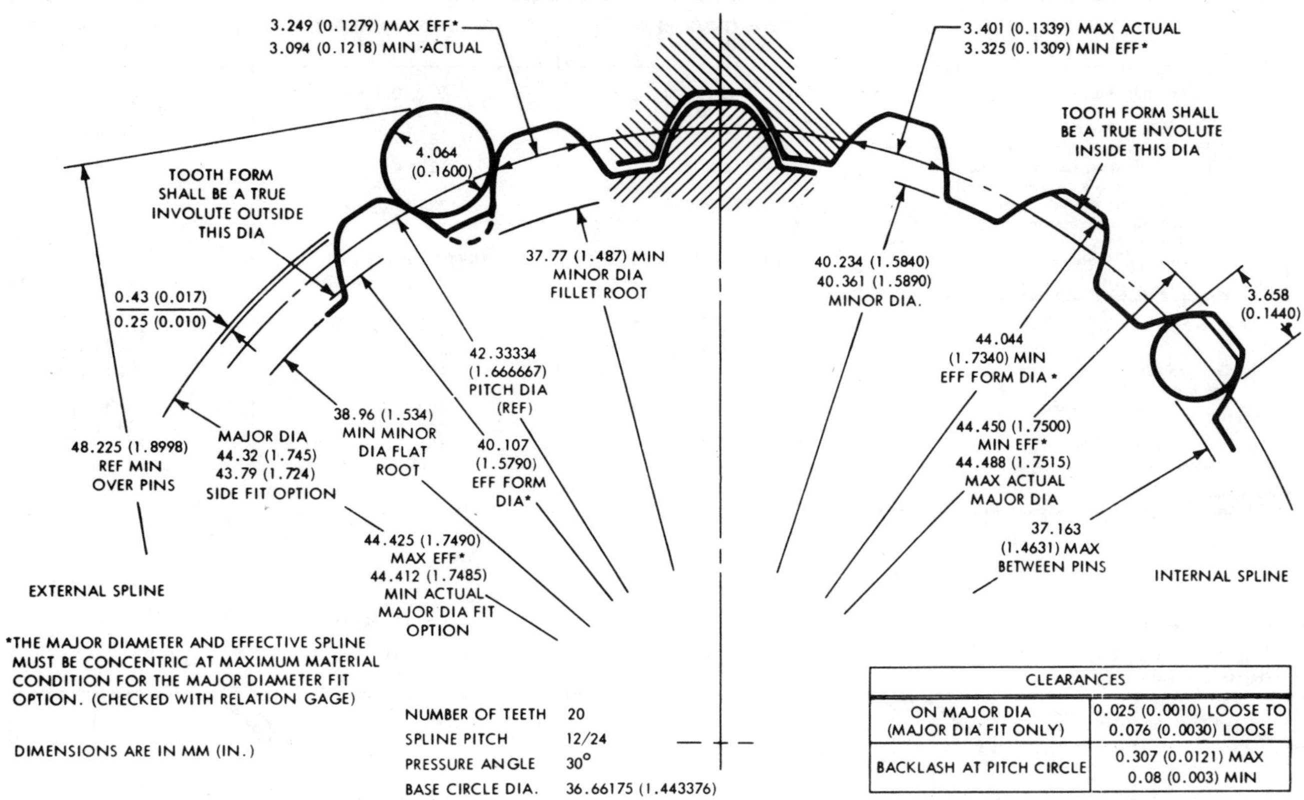

FIG. 4—1000 R/MIN POWER TAKE-OFF (45 mm ($1\frac{3}{4}$ in) INVOLUTE SPLINE DIMENSIONS)

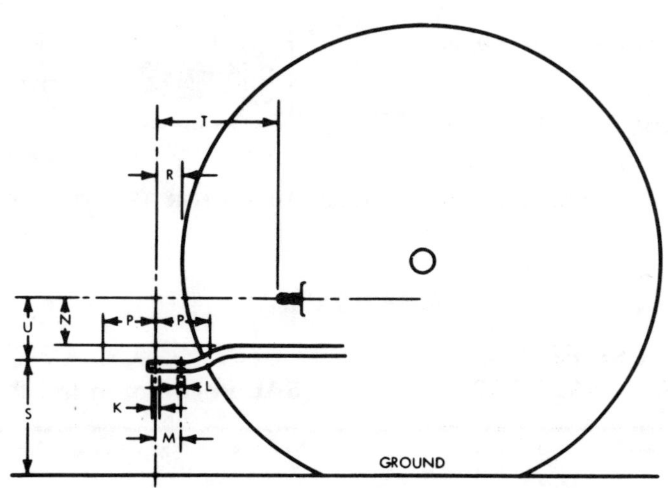

FIG. 5—TRACTOR POWER TAKE-OFF AND DRAWBAR (SEE TABLE 2)

TABLE 2—DIMENSIONS ASSOCIATED WITH DRAWBAR AND POWER TAKE-OFF[a] (SEE FIG. 5)

		35 mm (1-3/8 in) Dia. 540 r/min	35 mm (1-3/8 in) Dia. 1000 r/min	45 mm (1-3/4 in) Dia. 1000 r/min
K	Hitch pin hole dia., min	20.6 (0.81)	20.6 (0.81)	33.3 (1.31)
L	Auxiliary hole dia.	17.3 (0.68)	17.3 (0.68)	17.3 (0.68) min
M	Auxiliary hole spacing	102 (4.0)	102 (4.0)	102 (4.0)
N	Hitch pin and clevis clearance plane below PTO shaft centerline, min	122 (4.8)	122 (4.8)	165 (6.5)
P	Hitch pin and clevis clearance plane from hitch pin centerline, min	203 (8.0)	203 (8.0)	203 (8.0)
R	[b]Horizontal distance from hitch pin hole to tire:			
	OD Preferred	25 (1.0) to 127 (5.0)	25 (1.0) to 127 (5.0)	25 (1.0) to 127 (5.0)
	Max	−25 (−1.0) to 127 (5.0)	−25 (−1.0) to 127 (5.0)	−25 (−1.0) to 127 (5.0)
S	Height of drawbar with popular sized tire:			
	Preferred	381 (15.0)	381 (15.0)	483 (19.0)
	Min	330 (13.0)	330 (13.0)	432 (17.0)
	Max	432 (17.0)	432 (17.0)	533 (21.0)
T	End of PTO shaft to hitch pin hole	356 (14.0)	406 (16.0)	500.8 (20.0)
U	Top drawbar to PTO centerline:			
	Preferred	203 (8.0)	203 (8.0)	229 (9.0)
	Min	152 (6.0)	152 (6.0)	203 (8.0)
	Max	305 (12.0)	305 (12.0)	254 (10.0)

[a]Dimensions are in mm (in).
[b]Largest code R1 tire specified for use by the tractor manufacturer.

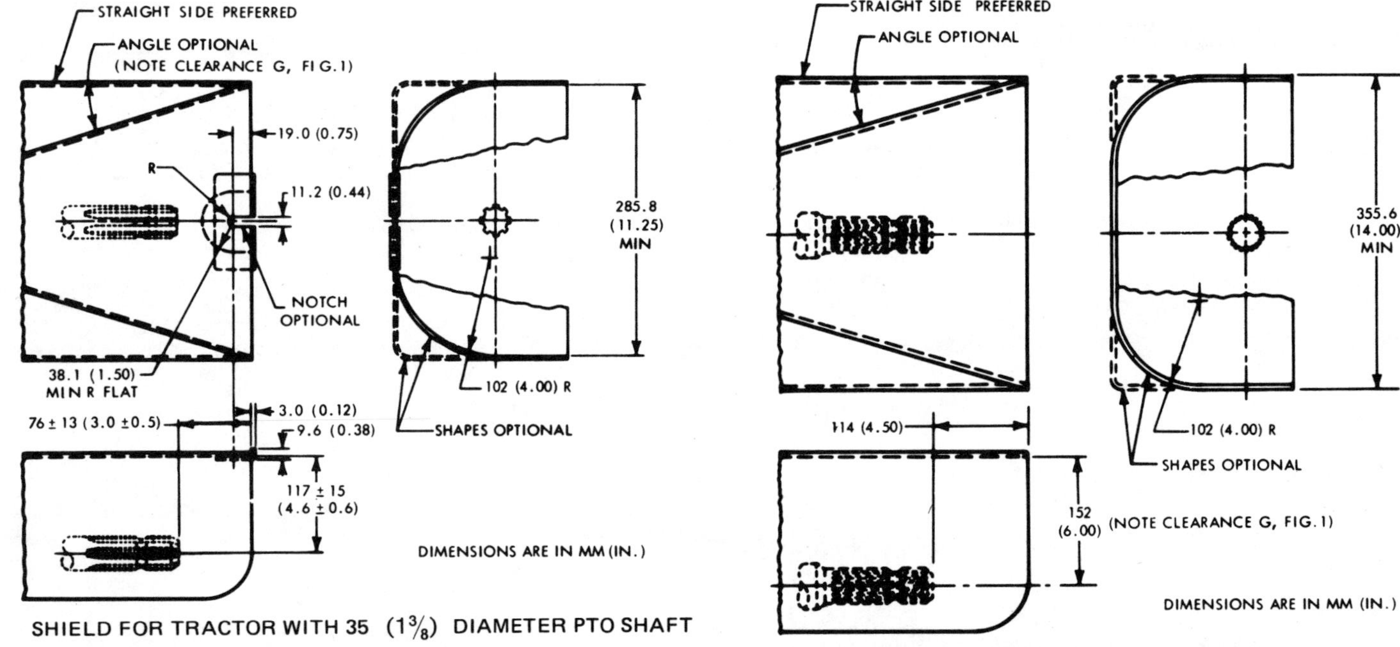

FIG. 6—POWER TAKE-OFF MASTER SHIELD FOR THE TRACTOR

APPENDIX B

AUXILIARY POWER TAKE-OFF DRIVES FOR AGRICULTURAL TRACTORS—SAE J717c

SAE Recommended Practice

Report of Tractor Technical Committee approved June 1955 and last revised June 1972. Conforms to report of FEI Advisory Engineering Committee, Chicago, Illinois. Editorial change June 1977.

1. Purpose and Scope—This document establishes specifications for mid and side power take-off drives that will be helpful in designing implements which are front or side mounted. Design of the implement attaching means must be tailored to each tractor, depending on attaching points available and the exact location of the auxiliary shaft.

2. Specifications

2.1 The auxiliary power take-off shafts shall conform to the dimensions of the 1$^3/_8$ in (34.9 mm) involute power take-off spline as specified in SAE J1170.

2.2 The normal speed of the auxiliary power take-off shaft shall conform to SAE J1170.

2.3 A minimum spherical clearance radius of 3.25 in (82.6 mm) with sufficient openings to allow for equipment drives shall be provided as specified for the 1$^3/_8$ in (34.9 mm) involute power take-off spline in SAE J1170.

2.4 The ability of the auxiliary power take-off shaft to withstand side loading and torque shall conform to SAE J721.

2.5 Machines driven from the auxiliary power take-off shall be equipped with adequate shielding for that portion of the power line that is furnished as a part of the driven machine. This shielding shall prevent the operator from coming in contact with the positively driven members of the power line. Provisions shall be made so that the power take-off shaft is covered when not connected to a driven assembly.

2.6 Position, location, and direction of rotation of the auxiliary power take-off shaft are shown in Table 1.

TABLE 1

	Mid	Right	Left
Position	Extends forward parallel to rear power take-off shaft	Right side of tractor extending to right when viewed from driver's station	Left side of tractor extending to left when viewed from driver's station
Location	Within zone extending 10 in. (254 mm) to both the right and the left of the centerline of the tractor	No recommendation	No recommendation
Direction of Rotation	Clockwise (when facing direction of forward travel)	Clockwise (when viewed from outer end of shaft)	Counterclockwise (when viewed from outer end of shaft)

APPENDIX C

OPERATING REQUIREMENTS FOR POWER TAKE-OFF DRIVES—SAE J721f

SAE Recommended Practice

Report of Tractor Technical Committee approved June 1953 and last revised June 1975. Conforms to report of FIEI Advisory Engineering Committee, Chicago, Illinois. Editorial change June 1977.

1. Purpose and Scope

1.1 This SAE Recommended Practice was prepared to assist manufacturers of tractors and pto-driven machines in providing suitable means of power transmission from the tractor to the driven machine.

1.2 SAE J1170 specifies the essential dimensions of tractor components necessary to enable manufacturers of driven machines to provide for the satisfactory hitching of their machines to any make of tractor. The successful performance of all tractor and driven-machine combinations likely to be met in field service requires consideration of many factors other than the dimensional relationships established in the aforementioned SAE Standard. Some of the more important of these factors are as follows.

2. Instructions for the Operator—

2.1 The tractor manufacturer shall provide a safety instruction in a prominent place on the tractor specifying:

2.1.1 The normal operating speed of the rear power take-off shaft (540 or 1000 rpm).

2.1.2 That the tractor drawbar is to be adjusted and locked in a position so that the drawbar hitch point is located in accordance with dimensions shown in SAE J1170.

2.1.3 That powerline safety shields are to be kept in place.

2.2 The implement manufacturer shall provide an instruction in a prominent place on the implement specifying:

2.2.1 The normal operating speed of the power take-off drive to the implement (540 or 1000 rpm).

2.2.2 That the powerline safety shields are to be kept in place.

2.3 The operator's manuals for both tractors and power take-off-driven implements shall also include the preceding information.

2.4 If a conversion assembly is made available for changing tractors or implements from the 540 to the 1000 rpm power take-off standard, or from the 1000 to the 540 rpm power take-off standard, these conversion assemblies shall include an instruction plate or sticker specifying the power take-off speed and the corresponding drawbar adjustments.

3. Implement Hitch and Powerline Design Requirements—

3.1 The hitch and powerline of any pto-driven machine, when the machine is hitched to any tractor that conforms to SAE J1170 should provide satisfactory operation over any terrain the machine is likely to encounter. To meet any such operating conditions, provisions should be made in the powerline and hitch of the driven machine to prevent any of the following from occurring:

3.1.1 The universal joints in the powerline from reaching a locking angle.

3.1.2 The telescoping section of the powerline from separating beyond the point where there is sufficient bearing to provide for proper operation.

3.1.3 The telescoping of the powerline from shortening to a solid position.

3.2 In normal forward operation the universal joints in the powerline of the driven machine should be in straight alignment as nearly as possible, and should be positively indexed with respect to each other so as to maintain the torsional-load fluctuations at the lowest possible value. Extreme care should be taken to determine load fluctuations or load reversals when employing one or three universal joints in a powerline.

3.3 Vertical Drawbar Loads

3.3.1 The minimum vertical static loads which the tractor drawbar must withstand are shown in Table 1.

3.3.2 The maximum vertical static loads which the equipment shall impose upon the tractor drawbar are shown in Table 1. The dynamic loads imposed upon the tractor drawbar and equipment hitch, at these static load ratings, will be considerably higher.

4. Maximum Bending Load Limitations for Power Take-Off Shaft Drives Employing V-Belts or Chains—

4.1 The power take-off drive of tractors is designed primarily to transmit torsional loads. When V-belt or chain drives with the driving sheave or sprocket mounted directly on the power take-off shaft are used, bending loads on the shaft should be checked carefully. The total bending load imposed on the tractor power take-off shaft by drives of this type should not be in excess of values shown in the table.

Position of Load Application	1-3/8 Dia Power Take Off		1-3/4 Dia Power Take Off	
	lb	kg	lb	kg
At end of power take-off shaft	500	227	800	363
Between power take-off shaft rear bearing and/or at groove in outside diameter of power take-off shaft splines	600	272	1000	454

TABLE 1—VERTICAL STATIC LOADS

Max HP[a]	Drawbar Load
20–100	750 lb plus 25 lb/HP for excess over 20 HP
100–250	2750 lb plus 10 lb/HP for excess over 100 HP
250–500	4250 lb plus 5 lb/HP for excess over 250 HP

[a] "Maximum drawbar horsepower established per SAE Standard J708, Agricultural Tractor Test Code, Maximum Drawbar Power with Ballast."

The tractor power take-off shaft and bearing mountings should successfully withstand the magnitude of bending loads shown in the table.

5. Powerline Protective Couplings and Maximum Torsional Load Limitations for Power Take-Off Shafts—

5.1 The dynamic torsional loads on power take-off drives should be checked carefully. Because of the large amount of kinetic energy available at the power take-off shaft, instantaneous torsional loads and fluctuating operating loads far in excess of the average rated horsepower of the tractor may be transmitted. Transmittal of these excessive loads can result in premature failure of the driving parts.

5.2 Implements subject to high starting loads or plugging should be equipped with an overload protective device in the powerline which will protect the drive against torsional overloads of sufficient magnitude to cause mechanical failure of either tractor or implement parts.

5.3 In consideration of the foregoing factors it is desirable for implements to conform to the following conditions:

5.3.1 The instantaneous operating loads should not exceed 9000 lb-in. (103.7 kg-m) for the $1^3/_8$ diameter shaft or 15,000 lb-in. (172.8 kg-m) for the $1^3/_4$ diameter shaft under conditions where there is no reversal of load. When a repetitive reversal of load is encountered, the aforementioned load limitations must be reduced by the amount of the reverse load. When the frequency of the instantaneous load does not exceed 10 cycles/hr, implements imposing loads greater than 9000 lb-in. (103.7 kg-m) for the $1^3/_8$ diameter shaft or 15,000 lb-in. (172.8 kg-m) for the $1^3/_4$ diameter shaft should have a powerline protective device which does not exceed a maximum instantaneous slip value of 15,000 lb-in. (172.8 kg-m) for the $1^3/_8$ diameter shaft or 26,000 lb-in. (299.6 kg-m) for the $1^3/_4$ diameter shaft.

5.3.2 The requirements in paragraph 5.3.1 will generally be met with a smooth-surface frictional type of powerline protective device which does not exceed a breakaway value of 4000 lb-in. (46.1 kg-m) for the $1^3/_8$ diameter shaft or 6400 lb-in. (73.7 kg-m) for the $1^3/_4$ diameter shaft when checked under static conditions. Snap-type spring-loaded jaw clutches should not be evaluated under static conditions, and, therefore, must comply with paragraph 5.3.1 above.

5.3.3 Power take-off implements which are capable of being loaded with a continuous torque in excess of 3000 lb-in. (34.6 kg-m) at the drive shaft shall be equipped with a low friction telescoping shaft incapable of transmitting an axial thrust in excess of 1500 lb (680 kg).

5.3.4 Tractors capable of imposing inertia loads on the powerline as high as those previously mentioned should have a power take-off drive capable of transmitting a torque equivalent to that described in paragraph 5.3.1 without failure. Tractors which are not capable of transmitting a torque of this magnitude, should have a drive of sufficient strength to transmit the maximum torque they are capable of delivering to the power take-off drive.

APPENDIX D

POWER TAKE-OFF DEFINITIONS AND TERMINOLOGY FOR AGRICULTURAL TRACTORS—SAE J722b

SAE Recommended Practice

Report of Tractor Technical Committee approved November 1954 and last revised June 1972.

1. Purpose and Scope

1.1 This SAE Recommended Practice deals only with the definitions and terminology pertaining to the power shafts of agricultural tractors in which the power take-off rotational speed is proportional to the engine speed. This is the type that prevails in the United States, Canada, England and, generally, throughout the world. The following recommendations will facilitate a clear understanding for engineering discussions, comparisons, and the preparation of technical papers.

2. Terminology

2.1 Clutch, Master—The term "master clutch" is generally used to describe a clutch which transmits all power from the engine and controls both travel and the power take-off. Likewise, when disengaged, both stop.

2.2 Power Take-Off, Transmission Driven—Power to operate both the transmission and the power take-off is transmitted through a master clutch, which serves primarily as a traction clutch. The power take-off operates only when the master clutch is engaged. The transmission driven power take-off ceases to operate at any time the master clutch is disengaged.

2.3 Power Take-Off, Continuous Running—Power to operate both the transmission and the power take-off is transmitted through a master clutch. Both operate only when the master clutch is engaged. Auxiliary means are provided for stopping the travel of the tractor without stopping the power take-off. The continuous running power take-off ceases to operate at any time the master clutch is disengaged.

2.4 Power Take-Off, Independent—Power to operate the transmission and power take-off is transmitted through independent transmission and power take-off clutches. Travel of the tractor may be started or stopped by operation of the transmission clutch without affecting operation of the independent power take-off. Likewise, the power take-off may be started or stopped by the power take-off clutch without affecting tractor travel.

APPENDIX E

TRACTOR POWER TAKE-OFF

S. A. E.—A. S. A. E. Standard

The tractor power take-off specifications as originally adopted by the S. A. E. in 1924, employed the S. A. E. Standard 6B spline for two shaft sizes and included the 536 r.p.m. take-off shaft speed. Subsequently the American Society of Agricultural Engineers adopted the standard but included an intermediate shaft size, additional data as to the location of the take-off on the tractor with respect to the drawbar and the power driven implement, and other requirements as to the throw-out clutch, supplementary equipment, etc.

As this specification is in the fields of both Societies[1] it was considered desirable that the standard as published by both should be the same, and accordingly in 1937 the following was recommended by the Agricultural Power Equipment Division of the S. A. E. Standards Committee and adopted by the Society in January 1938 to supersede the previous standard. (*Jan. '38*).

[1] By general understanding the standardizing activities of the S. A. E. in the agricultural engineering field are primarily concerned with the power supplying units, while those of the A. S. A. E. are primarily concerned with the units using power and the application of the power. This standard therefore lies in the juncture of these two fields of interest.—Ed.

The *tractor power take-off shaft* (except belt pulley shafts) for all types of power take-off drives issuing from the tractor in any direction shall have the following spline dimensions, and a minimum clearance of $3\frac{1}{4}$ in. between the end of the spline shaft and any stationary part of the tractor, this clearance to be the spherical radius from the center of the spline shaft end.

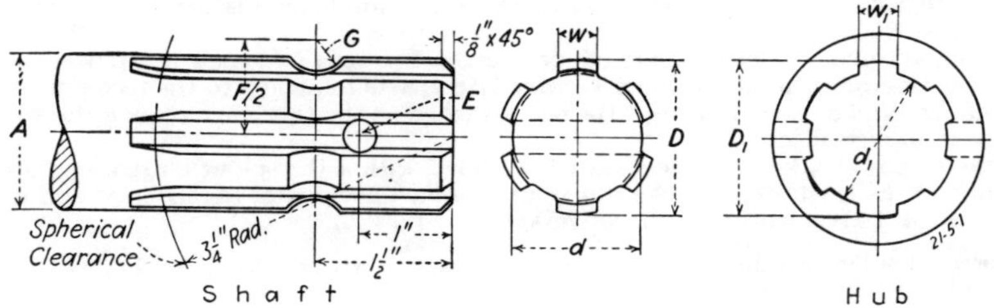

TABLE 1—SPLINE SHAFT AND HUB DIMENSIONS

A Nom. Diam.	Spline on Shaft						Spline in Hub						Radius R Max.
	D Large Diam.		d Small Diam.		W Spline Width		D_1 Large Diam.		d_1 Small Diam.		W_1 Spline Width		
	Max.	Min.	Max.	Min.	Max.	Min.	Max.	Min.	Max.	Min.	Max.	Min.	
1⅛	1.123	1.121	0.932	0.922	0.277	0.275	1.125	1.124	0.957	0.955	0.281	0.279	0.010
1⅜	1.373	1.371	1.145	1.135	0.340	0.338	1.375	1.374	1.170	1.168	0.344	0.342	0.010
1¾	1.747	1.745	1.454	1.444	0.433	0.431	1.750	1.749	1.489	1.487	0.438	0.436	0.015

TABLE 2—GENERAL SHAFT AND HUB DIMENSIONS

Shaft Diam. A	Shaft					Hub Pin Hole H
	Hole E	Pin Diam.	F	Side Rad. G	Groove Bottom Diam. Min.	
1⅛	13/64	3/16	1 15/32	17/64	15/16	13/64 (0.203)
1⅜	21/64	5/16	1 11/16	17/64	1 5/32	21/64 (0.328)
1¾	25/64	3/8	2⅛	21/64	1 15/32	25/64 (0.390)

The spline shaft groove is provided for use where additional retaining means is desired.

Power Take-off Speed

The normal speed of the power take-off shaft shall be 536 rpm, ±10 rpm—the direction of rotation to be clockwise when facing in the direction in which the tractor travels. (This applies only to fore-and-aft power take-off shafts.)

General Recommendations for Power Take-offs

(S. A. E.-A. S. A. E. Recommended Practice)

The 1⅜ in. shaft is recommended as desirable for future design, where practicable.

The end of the power take-off shaft on the tractor shall be 25 in., ±5 in., above the ground line. (Measurements to be taken without wheel lugs.)

The power take-off shaft, or an extension thereof, shall extend far enough to the rear of the tractor to clear the fenders and platform, and the hitch point of the drawbar shall be 10 to 15 in. back from the end of the splined portion of the shaft.

The horizontal location of the power take-off shaft shall be as near the tractor center line as possible with a tolerance of 5 in. right or left of center.

The telescoping members of the power shaft shall be so arranged that the tractor universal joint and the adjacent implement universal joint cannot be placed in improper relationship. (The tractor universal joint is in correct relationship to the adjacent implement universal joint when the two shaft yokes are in the same plane.) Slip clutches, if placed between these joints, should be designed to obviate improper alignment.

In plan view the tractor drawbar hitch point should be located midway between pivotal centers of the tractor universal joint and the adjacent driven implement universal joint (where the driven implement is being driven in normal working position) to equalize universal joint angularity when turning.

In side elevation the tractor drawbar hitch should be as close as practicable to the drive shaft connecting the tractor universal joint and the adjacent driven implement universal joint, to reduce the telescoping of the drive shaft to a minimum in traveling over rough ground.

The tractor manufacturer shall furnish the standard power take-off shaft.

The tractor manufacturer shall adequately shield the power take-off shaft and tractor universal joint, and provide protection for the operator against the telescoping member attached thereto, assuming connection between tractor and implement is according to recommended practice.

The manufacturer of a power take-off driven machine shall furnish the power drive parts up to the tractor spline shaft, the necessary hitch parts to attach to the recommended drawbar location, and all shields, except the one attached to the tractor, covering the spline shaft fitting or universal joint.

The tractor power take-off drive shall be provided with a throw-out clutch, operating independently of the tractor travel, of a design safe against accidental engagement and with a control located conveniently to the operator.

From the report of the Agricultural Power Equipment Division adopted by the Society July 1923. Last revision by the Tractor Division January 1938.

APPENDIX F

TRACTOR DRAWBARS

INDUSTRIAL (TRACK TYPE) TRACTOR DRAWBARS

SAE Standard

This standard is intended primarily for tractors designed for industrial uses and includes only those features which will affect the attaching of towed implements. As much freedom of design as possible, including features such as over-all jaw height, pin length and locking devices, swing angle and method of locking the bar, is left to the tractor manufacturer.

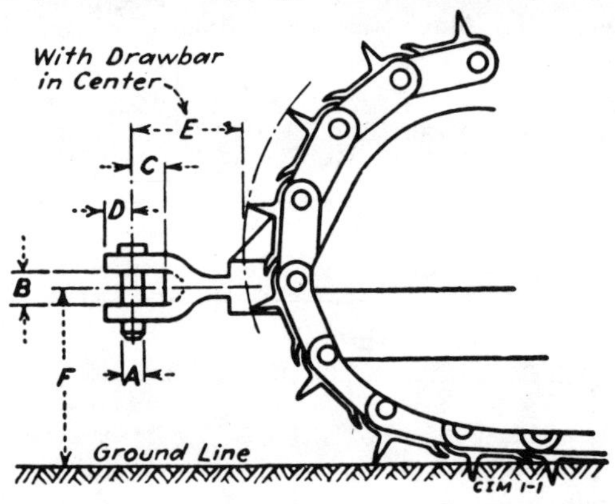

Tractor Size	Drawbar HP	A	B	C Min.	D Max.	E Min.	F	R
Class I	0– 50	1⅜	2½	2¼	2xR	4	12–15	
Class II	51–125	1¾	3½	3¼	3xR	6	14–18	
Class III	126–175	2	3¾	4	3½xR	8	16–21	

Report of the Construction and Industrial Machinery Technical Committee adopted December 1947.

AGRICULTURAL TRACTOR DRAWBARS

SAE—ASAE Standard

The first SAE specification for agricultural tractor drawbar location was adopted in 1917 shortly after the Society of Tractor Engineers amalgamated with the Society of Automobile Engineers. The standard, as revised in 1937, was adopted jointly by the American Society of Agricultural Engineers, and is accordingly published as SAE-ASAE Standard.

The diameter of the hitch hole at the end of the tractor drawbar shall be not less than 13/16 in. and in addition, an 11/16 in. hole shall be provided in the drawbar 4 in. ahead of the hitch hole.

The material in the tractor drawbar shall clear an implement clevis (3 in. wide and having a 3 in. throat clearance) through a 90 deg. swing right or left of the tractor drawbar center line.

The tractor drawbar shall be strong enough to carry a 500 lb. vertical load at the hitch point.

The following recommendations are intended to simplify the problem of connecting the power take-off of a driven machine to that of a tractor. When these standards are incorporated in the design of power take-off drives, any power take-off driven machine can be quickly connected to any make of tractor without the necessity of supplying special equipment.

The horizontal distance B between the hitch point on the tractor drawbar and the rear-most point on the standard size rubber tire, steel wheel rim lug or fender of the tractor shall be not less than 4 in.

The vertical distance C from the ground line to the top of the drawbar at the hitch point shall be from 12 in. to 15 in., the 15 in. distance being recommended practice.

AGRICULTURAL TRACTOR DRAWBARS

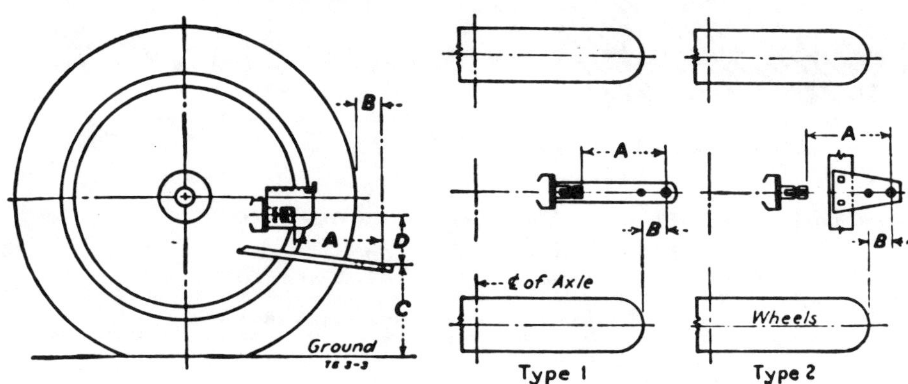

FIG. 1—POWER TAKE-OFF AND DRAWBAR HITCH LOCATIONS FOR AGRICULTURAL TRACTORS AND MACHINES

The horizontal distance A between the hitch point on the tractor drawbar and the end of the splined shaft of the power take-off shall be 14 in., and the hitch point shall be directly in line with the centerline of the power take-off shaft.

The vertical distance D between the top of the drawbar at the hitch point and the center line of the power take-off splined shaft shall be not less than 6 in. nor more than 15 in., 8 in. being the recommended practice.

The location of the tractor power take-off shaft shall be within the limits of 3 in. to the right or left of the center line of the tractor. The center line location of the power take-off shaft is recommended practice.

The weight imposed by the drawn machine or implement on the tractor drawbar at the hitch point shall not exceed 500 lbs.

Report of Agricultural Power Equipment Division adopted July 1923. Last revision by Tractor Division January 1946.

APPENDIX G

AGRICULTURAL TRACTOR POWER TAKE-OFF

SAE—ASAE Standard

The tractor power take-off specifications as originally adopted by the SAE in 1924, employed the SAE Standard 6B spline. Subsequently the American Society of Agricultural Engineers adopted the standard also and it is accordingly published as standard of both Societies.

The agricultural tractor power take-off shaft (except belt pulley shafts) for all types of power take-off drives issuing from the tractor in any direction shall have the following spline dimensions, and a minimum clearance of 3¼ in. between the end of the spline shaft and any stationary part of the tractor, this clearance to be the spherical radius from the center of the spline shaft end.

The spline shaft groove is provided for use where additional retaining means is desired.

The normal speed of the power take-off shaft shall be 536 rpm, ±10 rpm —the direction of rotation to be clockwise when facing in the direction of forward travel. (This applies only to fore-and-aft power take-off shafts.)

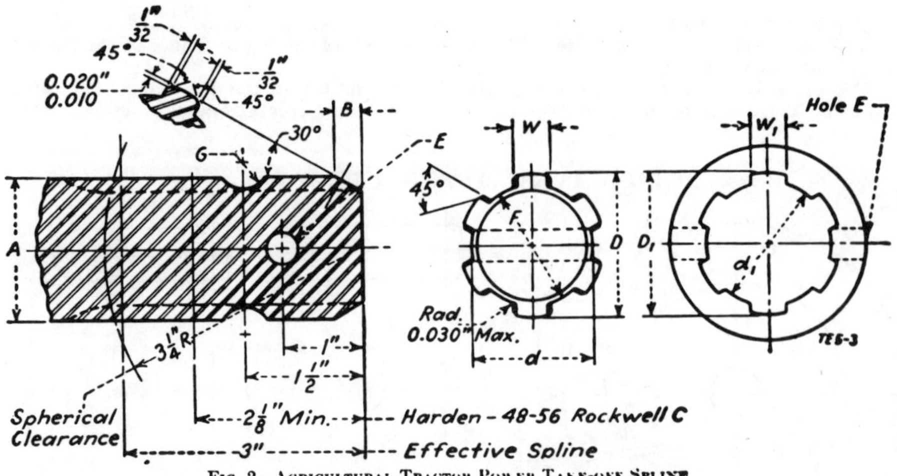

Fig. 2—Agricultural Tractor Power Take-off Spline

TABLE 1—SHAFT AND HUB SPLINE DIMENSIONS

A Nom. Diam.	Spline on Shaft						Spline in Hub					
	D Large Diam.		d Small Diam.		W Spline Width		D_1 Large Diam.		d_1 Small Diam.		W_1 Spline Width	
	Max.	Min.	Max.	Min.	Max.	Min.	Max.	Min.	Max.	Min.	Max.	Min.
1⅜	1.373	1.371	1.108	1.098	0.340	0.338	1.375	1.374	1.170	1.168	0.344	0.342
1¾	1.747	1.745	1.427	1.417	0.433	0.431	1.750	1.749	1.489	1.487	0.438	0.436

TABLE 2—GENERAL SHAFT AND HUB DIMENSIONS

A Shaft Diam.	Shaft						Hub
	E Pin Hole Diam.	Pin Diam.	B Chamfer Length	G Side Rad.	F Groove Bottom Diam., Min.		E Pin Hole Diam.
					Max.	Min.	
1⅜	21/64 (0.328)	5/16	9/32	17/64	1.160	1.155	21/64 (0.328)
1¾	25/64 (0.390)	3/8	11/32	21/64	1.470	1.465	25/64 (0.390)

Power Take-off Master Shield

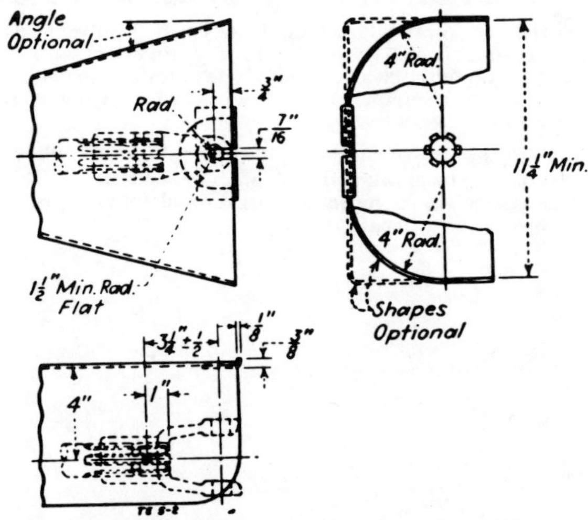

Fig. 3—Agricultural Tractor Power Take-off Master Shield

1. The tractor shall be equipped with a power take-off master shield, incorporating the attaching point for the shield of the driven machine as dimensioned.

2. The strength of the master shield shall be sufficient to support the weight of the operator without taking a permanent set.

3. The manufacturer of the driven machine shall provide adequate shielding for that part of the power line drive which he furnishes and which shall be attachable to the standard power shaft master shield of the tractor.

[1] *By general understanding the standardizing activities of the SAE in the agricultural engineering field are primarily concerned with the power supplying units, while those of the ASAE are primarily concerned with the units using power and the application of the power. This standard therefore lies in the juncture of these two fields of interest.—Ed.*

Report of Agricultural Power Equipment Division adopted July 1923. Last revision by Tractor Division January 1946.

APPENDIX H

FULL SHIELDING OF POWER DRIVELINES FOR AGRICULTURAL IMPLEMENTS AND TRACTORS — SAE J955a

SAE Recommended Practice

Report of Tractor Technical Committee approved June 1966 and last revised June 1967. Editorial change September 1968. Conforms to ASAE S297T.

1. Purpose and Scope—This standard establishes the specifications for a fully shielded power take-off power line for both towed and integral machines.

2. General

2.1 The shielding of the power driveline, including all yokes and joints, shall be adequate to prevent the operator from coming in contact with positive-driven rotating members of the power driveline. This protection shall be adequate during turns under load.

2.2 The shaft shield and yoke shield (Fig. 1) shall be integral with and journaled on the rotating members.

2.3 The shields shall be free from protuberances and shall cover the knuckles at any angle, α, assumed by the universal joint during operation and shall not present a pinch point during articulation. See Fig. 1.

2.4 A maximum spherical clearance radius (Fig. 1) shall be maintained as defined in SAE J718 and SAE J719.

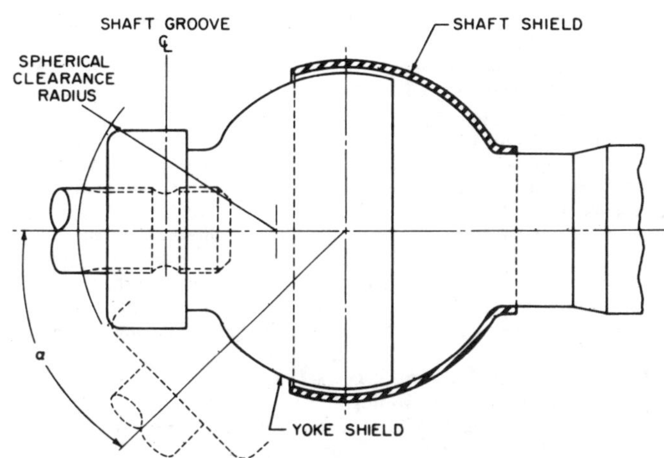

FIG. 1—FULL SHIELDING FOR POWER DRIVELINES

APPENDIX I

SAFETY FOR AGRICULTURAL EQUIPMENT—SAE J208d

SAE Standard

Report of Tractor Technical Committee approved July 1970 and last revised by Agricultural Tractor Technical Committee May 1978. Submitted by FIEI. Conforms to ASAE Standard S318.6. Rationale statement available.

1. Purpose and Scope

1.1 This standard is a guide to provide a reasonable degree of personal safety for operators and other persons during the normal operation and servicing of agricultural equipment.

2. Definitions

2.1 *Agricultural Equipment* means agricultural tractors, self-propelled machines, implements, and combinations thereof designed primarily for use in agricultural operations.

2.2 *Propelling Machine* means a tractor or self-propelled machine.

2.3 *A Shield (or Guard)* means a barrier which minimizes inadvertent personal contact with hazards created by moving machinery parts.

2.4 *Inadvertent Contact* means contact between a person and a moving machinery part hazard, or other type of hazard, resulting from the person's unplanned actions during normal operation or servicing.

2.5 *Moving Machinery Part Hazard* means a source of potential injury created by moving machinery parts which can cause serious injury upon direct contact or by entanglement of personal apparel. This includes but is not limited to the nip-points of power driven gears, belts and chains, and projections on rotating parts.

2.6 *Nip Point* means the pinch point of gears and the run-on point where a belt or chain contacts a sheave, sprocket, or idler.

2.7 *Guarded by Location* means a hazard is guarded when it is covered by other parts or components of the machine, or because of its remote location, inadvertent contact is minimized during normal operation or servicing.

2.8 *Power Take-Off (PTO)* means an external shaft on the rear of a tractor to provide rotational power to implements.

2.9 *Auxiliary Power Take-Off (Aux. PTO)* means an external shaft on a tractor, other than the rear PTO, to provide rotational power to implements that are usually front or side mounted.

2.10 *Implement Drive Line* means the shafts, universal joints, connectors, and fasteners provided with the implement to transmit rotational power from the tractor PTO to the first component on the implement, such as a gear set, pulley, sprocket, or flywheel.

2.11 *Ground Driven Components* means components which are powered by the turning motion of a wheel as the equipment travels over the ground.

2.12 Refer to ASAE Standard S390/SAE Standard J1150, Terminology for Agricultural Equipment, for additional definitions.

3. Operator's Manual

3.1 Operator's manuals shall be supplied with each piece of equipment.

3.2 Operator's manuals shall provide general safety instructions for normal operation and servicing of the equipment.

4. Operator Controls

4.1 Location and movement of operator controls shall be in accordance with ASAE Standard S335/SAE Standard J841, Operator Controls on Agricultural Equipment.

4.2 Symbols used to identify controls shall be in accordance with ASAE Standard S304/SAE Standard J389, Symbols for Operator Controls on Agricultural Equipment. Controls for functions that are not obvious should be identified.

4.3 Foot operated controls shall be adequate size, spacing, and appropriate configuration for proper operation. A slip-resistant means shall be provided to minimize the possibility of the operator's foot slipping off the controls.

4.4 Hand-operated controls shall be of appropriate configuration and size to permit adequate grasp and hand clearance throughout the operating range.

5. Operation and Servicing Provisions

5.1 A suitable station shall be provided for each person required for the operation of the equipment.

5.2 All equipment shall have steps and handholds, or other means to facilitate entry and exit from the operating positions.

5.3 If the mounting step(s), ladder(s), or handholds(s) of the propelling machines are made inaccessible by the installation of equipment, alternate facilities shall be provided.

5.4 Handholds, hand rails, guard rails, or barrier type safeguards shall be provided, if necessary, to minimize falling during normal operation or servicing, unless means are provided by other parts of the equipment.

5.4.1 Guard rails when provided shall have a top rail 900–1050 mm (35.4–41.3 in) above the working walkway or platform with a rail approximately midway between the platform and the top rail.

5.5 The height of the first step should not exceed 686 mm (27.0 in), preferably 550 mm (21.6 in). The vertical distance between the steps should not exceed 406 mm (16.0 in), preferably 300 mm (11.8 in), and adequate toe clearance should be provided for each step.

5.6 Steps and operator platforms shall have a slip-resistant surface.

5.7 Shielding shall be provided on the back of steps or ladders wherever a protruding hand or foot may contact a moving machine part hazard.

5.8 Glazing material, such as glass or plastic used in operator enclosures shall be in accordance with SAE Standard J674, Safety Glazing Materials—Motor Vehicles.

6. Power Take-Off and Implement Drive Line

6.1 The tractor shall be equipped with a PTO master shield when provided with a PTO shaft in accordance with ASAE Standard S203/SAE Standard J1170, Rear Power Take-Off for Agricultural Tractors. Although not intended as a step, the master shield shall not permanently deform if used as a step by a 120 kg (265 lb) operator.

6.2 Provisions shall be made on the tractor to shield the Aux. PTO when it is connected to an implement drive line.

6.3 Shields or other protective means shall be provided on the tractor for the PTO and Aux. PTO when they are not connected to an implement drive line.

6.4 PTO driven implements shall be equipped with shielding for the implement drive line.

6.5 PTO driven implements that require removal of the tractor master shield shall include comparable shielding.

6.6 A safety sign(s) shall be provided at a prominent location on the implement specifying the normal PTO operating speed and that implement drive line shields are to be kept in place.

6.7 A safety sign(s) shall be provided at a prominent location on the tractor specifying the normal PTO operating speed and that PTO shields are to be kept in place.

6.8 PTO driven equipment designed to operate in a stationary position, should be provided with a means to maintain the implement to tractor spacing to prevent separation of the power drive line.

7. Tractor Roll-Over Protection—Tractor Roll-Over Protection, if installed, shall be in accordance with ASAE Standard S383/SAE Standard J1194, Roll-Over Protective Structures (ROPS) for Wheeled Agricultural Tractors.

8. Shields (or Guards)

8.1 The following shall be shielded or shall be guarded by location to minimize inadvertent contact with hazards created by:

8.1.1 Moving traction elements.

8.1.2 Revolving engine components.

8.1.3 Nip-Points of exposed gears, belts, chain drives, and idlers.

8.1.4 Outside faces of pulleys, sheaves, sprockets, and gears on drives that rotate when the engine is running with all clutches disengaged.

8.1.5 Revolving parts with projections such as exposed bolts, keys, or set screws.

8.1.6 Revolving shafts, except smooth shaft ends protruding less than one-half the diameter of the shaft including its locking means.

8.1.7 Ground driven components, if operating personnel are exposed to them while the drives are in motion.

8.1.8 Surfaces hot enough to cause personal injury during normal operation or servicing.

8.2 Functional components, such as snapping or husking rolls, straw spreaders and choppers, cutterbars, flail rotors, rotary beaters, augers, feed rolls, rotary tillers, and similar units which must be exposed for proper function shall be shielded to the maximum extent permitted by the intended function of the component(s).

8.3 Equipment with access doors and shields which can be opened or removed while components continue to rotate after the power is disengaged, shall have in the immediate area:

(a) a readily visible evidence of rotation or audible indication of rotation and,

(b) a suitable safety sign.

8.4 Access doors and shields that must be opened for normal servicing, shall be easily opened and closed.

9. Lifted Units—A means shall be provided to protect against inadvertent dropping of lifted units which must be in a raised position for normal servicing or adjusting. In addition, the operator's manual should contain instructions to securely support or block components which are raised before servicing or adjusting.

10. Travel on Highways

10.1 Lighting and marking of equipment shall conform to ASAE Standard S279/SAE Standard J137, Lighting and Marking of Agricultural Equipment on Highways.

10.1.1 The operator's manual for the unit shall instruct the operator to turn on the flashing warning lights whenever traveling on a highway except where such use is prohibited by law.

10.2 Propelling machines with operator enclosures shall have at least one rear-view mirror to permit the operator to see the highway behind the machine.

10.3 Hitch pins and other hitching devices shall be provided with a retainer to prevent accidental unhitching.

10.4 Components that are retracted to decrease the width for transport shall have means for securement during transport.

10.5 Provisions shall be made for the use of auxiliary attaching systems per the ASAE Recommendation R338, Safety Chain for Towed Equipment on equipment which is towed on highways by single point attachment.

11. Braking and Parking Requirements

11.1 For guidance in determining performance needs of agricultural equipment braking system refer to ASAE Standard S365T/SAE Standard J1041, Brake Test Procedures and Brake Performance Criteria for Agricultural Equipment.

11.2 All towed equipment with a tongue imposing a vertical downward force at the hitch point of more than 245 N (55 lb) at a height of 406 mm (16 in) when on level ground and any condition of loading shall be equipped with a means for attaching to the propelling machine without manual lifting.

12. Fire Protection

12.1 Shields shall be provided for the engine exhaust manifolds, muffler, and exhaust pipe when necessary to prevent contact with flammable crop materials.

12.2 Fuel sediment bowl assemblies used on gasoline engines shall be fire resistant.

13. Safety Signs

13.1 Safety signs shall be appropriately displayed when necessary to alert the operator and others of the risk of personal injury during normal operations and servicing.

13.2 Description

13.2.1. The signal words CAUTION, WARNING, DANGER are recommended to indicate the degree of hazard and are to be used for items of personal safety. (Not to be used for instructional signs relating to equipment servicing and care. These should use words such as—Important, Attention, or Notice.)

13.2.1.1 CAUTION is used for general reminders of good safety practices or to direct attention to unsafe practices.

13.2.1.2 WARNING denotes a specific potential hazard. The sign should be distinctively displayed in the area of the hazard.

13.2.1.3 DANGER denotes the most serious specific potential hazard. The sign should be distinctively displayed in the area of the hazard.

13.2.2 The *Safety Alert Symbol,* ASAE Standard S350/SAE Standard J284, Safety Alert Symbol for Agricultural Equipment, shall be used with each of the signal words to indicate that safety is involved.

13.3 Colors

13.3.1 CAUTION and WARNING signs shall have the color combination of yellow and black.

13.3.2 DANGER signs shall have the color combination of red and white.

13.3.3 A margin may be applied to any portion of the sign to achieve distinctiveness.

13.3.4 These color combinations should not be used for other informational signs.

13.4 Other Requirements

13.4.1 Safety signs, statements, or similar instructions shall be included in the operator's manual.

13.4.2 Safety signs shall be of high quality material. Safety signs shall withstand weather exposure at 45 deg upward and facing south in the Miami, FL area for 24 months without showing appreciable checking, chalking, dulling, color change, blistering, or loss of adhesion after exposure.

FIG. 1—TYPICAL SAFETY SIGNS WITH SAFETY-ALERT SYMBOL AND SIGNAL WORDS

800944

Milestones in the Application of Power to Agricultural Machines

Richard N. Coleman
Retiree of International Harvester

Keith W. Burnham
International Harvester

THIS PAPER WILL CONCENTRATE on the period from 1905 to 1980 for two reasons: first, as an honor to SAE on its 75th anniversary, and secondly, because most of the application of mechanical power to agriculture has taken place during this period.

Animal power was first probably used in agriculture through the use of the yoke and oxen about 2000-1500 B.C. (1)(2)*. Not much more use was made of power to agriculture other than human power until about 1000 A.D., when the horse collar was probably invented.

From the period of 1800 to 1850 or 1860, many inventions were recorded that permitted better use of animal power, such as the iron plow, the reaper, the steel plow, the thresher, and many others too numerous to mention. Mechanical power as derived from the steam engine originally was applied to agriculture as stationary power for threshing machines in the mid-1800's. The J. I. Case Company was a large producer of steam engines for agricultural applications. Later on, about 1870 or 1880, many of these steam engines were equipped with traction drives and were used for basic tillage, particularly in the west (Fig. 1).

The first use of internal combustion engines, such as gasoline, was recorded in the 1890's, basically for experimentation and trial. John Froelich is often credited with making the first successful gasoline tractor, about 1892, that would operate a significant period of time without failure (3).

J. I. Case, however, was also experimenting with gasoline engines in chassis that had previously had steam engines. So they were one of the early-birds in this business.

John Froelich started producing tractors for sale, as we understand it, in about 1902; and his facilities were ultimately purchased by Deere and Company.

Two young engineers by the names of Hart and Parr produced a gasoline tractor in 1902. White Farm Equipment Company can be traced back to this effort. Hart-Parr is credited with using the name "tractor" in place of "gasoline traction engine" in sales promotion in about 1906. Hart-Parr is also credited with having the first factory devoted

*Numbers in parentheses designate References at end of paper.

ABSTRACT

This paper lists and discusses the important milestones, as judged by the authors, in the application of internal combustion engine power to agricultural machines, particularly tractors. An evaluation of the efficiency of fuel use and the productivity of farm workers is included. Where possible the individual or company responsible for first introducing the selected milestone is cited.

SAE/SP-80/470/$02.50

Fig. 1 - Case steam tractor (J. I. Case Company), about 1905

entirely to tractor manufacture. This might be considered the first milestone for the period 1905 to 1980 in the application of power to agricultural machinery.

Using this as Number 1, we might list other major milestones thusly:

2. First frameless tractor, about 1912 or 1913, the Wallis Cub (part of the Massey-Ferguson family tree)(4).

3. The light weight, low cost Fordson, which sold thousands and thousands, about 1917 or 1918, Ford Motor Company (4).

4. The first power take-off for driving equipment remote from the tractor, about 1918, International Harvester (3)(4).

5. The first successful or widely accepted general purpose tractor, about 1924, International Harvester "Farmall" (3)(4).

6. The first tractor with a diesel engine, about 1931, Caterpillar Tractor Company (3)(4).

7. First use of rubber tires on agricultural tractors, about 1932, Farm Equipment Division, Allis-Chalmers Manufacturing Company and Firestone Tire & Rubber (3)(4).

8. First three-point hitch with hydraulic draft control, Harry Ferguson inventor, David Brown Tractor Company in Britain, about 1935; Ford Motor Company, about 1939, USA (5).

9. Self-propelled combine, about 1938, Massey-Harris (now Massey-Ferguson).

10. First marketing of LPG engine in agricultural tractor, about 1941, Minneapolis-Moline (now part of White Farm Equipment) (3)(4).

11. First successful cotton picker, about 1942, International Harvester. Strictly speaking, this machine was mounted on a modified tractor but it is difficult to visualize it as a PTO driven machine.

12. First continuous running PTO, about 1948, Cockshutt Tractor (now part of White Farm Equipment) (3).

13. First partial range power shift transmission, about 1953, Farmall Super M-TA, International Harvester. Minneapolis-Moline had a similar transmission at about the same time (3).

14. First full range power shift for agricultural tractor, Ford Selec-O-Speed, about 1957, Ford Tractor Operations (3).

15. First use of turbo-charged diesel engine in tractors specifically for agriculture, Allis-Chalmers, about 1962 (6).

16. First factory-supplied roll over protection, about 1965, Deere and Company.

17. First significant volume production and sale of articulated four-wheel drive tractors for agricultural use, about 1965, Versatile Manufacturing Company.

18. Wide spread use of factory-mounted cabs for tractors, about 1970 to 1972. It is difficult to give credit to any one company as being first. Certainly independent cab manufacturers had a lot to do with this. Also, Minneaplis-Moline (now White Farm Equipment Company) produced a tractor with an all-weather cab way back in about 1938.

These are only the major milestones; of course, many other things were occurring and are deserving of some discussion. For example, in the early 1900's, between 1905 and 1910, kerosene became the major fuel for agricultural tractors. On most of these units, water was fed into the induction system along with the fuel to control detonation and knocking under heavy load and high temperature operating conditions. We often forget that agricultural equipment manufacturers had much experience with water injection many years ago.

During this 1905 to 1910 period, the gas engine or internal combustion engine had not really taken over from the steam engine as a source of power for tractor operation. In Winnipeg, Canada, in the years 1908 through 1912, in connection with the Winnipeg Industrial Exhibition, competitive demonstrations were made of various tractors from a performance and field operation point of view. These contests proved that the tractors powered by internal combustion engines were more efficient than those powered by steam. A review of the contest results shows that in 1908 most of the internal combustion engine tractors were operated on gasoline. By 1911 and 1912, however, a significant number of machines were being operated on kerosene. The power per unit of volume of fuel was about equal whether gasoline or kerosene was used. Kerosene was being used on farms to heat, to cook, for chicken brooders, and so forth. It was only natural that it would be an attractive fuel for the farmer since gasoline was in short supply and considerably more expensive. Kerosene was used as a fuel in 53 of the 65 tractors tested at Nebraska in 1920, the first year of Nebraska Test. Gasoline was used in the other twelve. The last tractor to burn kerosene during Nebraska Test was in 1934. For typical tractors of the 1905-1915 period, see (Figs. 2, 3 and 4).

Many of the first engines to be put in tractors were taken from stationary applications. They had open crankcases, hopper cooling, and the transmissions and drives on the tractors were often also open. In the period 1905 to 1915, things changed rapidly. Gearings were enclosed, crankcases were enclosed, and it soon became evident that something was going to have to be done about the dirt that was being drawn into the engine during field operations.

There were a great many small manufacturers in those days; tractors came in all sizes, shapes, and configurations: tricycle, two-wheel drive, single-wheel drive, front-wheel drive, crawler--you name it, somebody had it. In fact, there is even an account of one with power steering; however, we're sure it was not hydraulic power steering. During the mid-teens, when the first frameless tractor made its appearance (listed as Number 2 in our milestones), light weight tractors became more and more popular. In 1917, stimulated by a huge order from Britain, because of the war, Ford Motor Company introduced the Fordson, a light weight, low cost tractor (Fig. 5). It was introduced to the USA in 1918. By the early 20's, this tractor had a giant share of the total tractor market. The Fordson was of unit-frame construction, only its frame was made of cast iron--a first. It also had an air cleaner in which water was used as a cleaning agent. It is our understanding that in cold weather kerosene or oil was substituted for water as a cleaning agent for the dust.

It was on the International 8-16 tractor

Fig. 2 - Hart-Parr Old Reliable, circa 1907-1913

Fig. 3 - Titan tractor (International Harvester Company), about 1910

Fig. 4 - Rumely Oil Pull (Advanced Rumely), about 1910

Fig. 5 - Fordson (Ford Motor Company), 1917

Fig. 6 - Farmall (International Harvester Company), about 1926

that the first power take-off for driving remote implements was introduced by International Harvester (Number 4 item in our list of major milestones). We won't discuss this further because the development of the power take-off is covered by another paper in this session: "The Development of Agricultural Equipment Power Take-Off Mechanism", T. H. Morrell.

It was in 1919 that the Nebraska law covering Nebraska Test was passed and in 1920 the first tests were run. This will not be covered much further in this paper because the tractor test code history and development is covered by another paper: "History and Development of Agricultural Tractor Test Code", Candee and Walters.

In 1922, Agricultural Engineering Department, University of California-Davis, under the direction of A. A. Hoffman, 26 air cleaners were tested and it was found that the efficiences varied from 42.7% to 99.8% (4). Engineers in agriculture must be given credit for much of the success of the present-day air cleaners used on practically all engine applications for land motive power.

Practically all historians writing about American agriculture mention the Farmall tractor as being the first successful general purpose tractor and one of the major milestones in the application of mechanical power to agriculture (Fig. 6). It was certainly not the first time that engine power had been used for cultivating. There were self-propelled cultivators prior to this; there were three-wheel tractors prior to this; power take-off units prior to this; but all of these factors were brought into one tractor, one machine that could be used to obtain a real gain in replacing animal power on the farm. In unpublished "Notes (1932 by International Harvester) on the Development of the Farmall Tractor" (7), the following statement is made: "It was a case of realizing a necessity for foreseeing a demand, not a case of developing something to meet a demand". The notes further indicate that the demand from the territory was: "Get us a tractor to compete with Fordson"; nothing was said about cultivating and all the other jobs that the Farmall tractor was capable of doing. It was pure "doggedness" on the part of the engineering department and Mr. Bert Benjamin that the Farmall tractor was ultimately placed in production. Demand soon became so great that a separate plant was obtained for its manufacture. The selling price was considerably more than the Fordson, so it had to compete on the basis it could do more things and better things than the Fordson or it would not have been a success.

In about 1928, Deere and Company, on their general purpose tractor, introduced a mechanical power lift as a tractor attachment (3). It was used for lifting mounted implements. In about the mid-30's, Deere replaced the mechanical lift with a hydraulic lift. Deere must be considered pioneers in this effort.

Caterpillar Tractor Company, in 1931, introduced the first diesel engine for use in agriculture in their 65 Caterpillar tractor. It certainly is a milestone, as practically all farm tractors today are diesel. The roots of Caterpillar go back to Benjamin Holt and a 1904 steam traction engine with tracks substituted for standard wheels (4).

In 1932, rubber tires were demanding attention even in the depths of the depression. Allis-Chalmers and Firestone Tire & Rubber are given credit for promoting the widest use of rubber tires on agricultural tractors (Fig. 7). Both the diesel engines and rubber tires were very significant items in making better use

Fig. 7 - Firestone and Allis-Chalmers, about 1932

of energy in agriculture. As we will see later in the paper, there has been a continued improvement in the use of energy throughout the entire 75-year period we are looking at.

It was previously mentioned that the last kerosene tractor was tested in Nebraska in 1934. During the mid-30's, three major fuels were being used in tractors: gasoline, distillate or tractor fuel, and diesel or fuel oil. It was about this time that minimum octane ratings were put on gasoline and compression ratios could be increased to give considerably better efficiency on gasoline engines. Distillate or tractor fuel took over where kerosene had been used previously. Most of the distillate burning engines did not require water injection to control detonation. Compression ratio was kept low enough and the octane rating of the distillate tractor fuel was improved to the extent where it was not really necessary.

In about 1938, Massey-Harris introduced the self-propelled combine. A very successful PTO driven combine had been introduced by Allis-Chalmers previous to this. We have chosen not to discuss the PTO driven machines much in this paper as we have restricted our discussion to machines that contain the primary power source. Later, during the Second World War, Massey-Harris received special permission from the government to build some 400 self-propelled combines to aid in the production of food materials during the war.

The Ferguson three-point hitch with draft control was introduced by Ford Motor Company in this country in 1939 on the Ford-Ferguson 9N (6) (Fig. 8). Ferguson had some exposure in Britain prior to this on a tractor built by David Brown. History has certainly shown three-point hitch and draft control is desired.

Fig. 8 - Ford 9N, 1939

Practically every row crop tractor today can be so equipped. It is our judgment the popularity is because the system permitted mounting tools without gage wheels so that all the weight of the tools and all the vertical soil forces could be added to the tractor to aid in traction and tractive efficiency. Three-point hitches for different size tractors have been standardized to the point where a three-point implement is interchangeable throughout most the world.

In 1941, Minneapolis-Moline introduced the first LPG--Liquified Petroleum Gas--tractor as a regular factory option. We are not aware of any LPG tractors available as factory equipment today. Beginning with Minneapolis-Moline's introduction in the early 40's until about mid-1960, LPG gained in popularity to the point where it became a significant part of tractor production and all major tractor manufacturers provided it as an option. It gradually lost favor after that time due to economic reasons.

Just as we highlighted the first use of self-propelled combines, we should also call attention to the first successful cotton picker about 1942 by International Harvester (Item 11 on list of major milestones). Cotton pickers have since become a major factor in reducing the tremendous amount of hand labor that went into the production of cotton.

Item 12 on our major list of milestones is the first continuous-running power take-off in 1946 by the Cockshutt Tractor Company, now part of White Farm Equipment. This deserves more discussion. Prior to this, power take-offs were driven from the transmission and whenever a transmission stopped turning, the power take-off stopped turning. In other words, when you disengaged the clutch to stop the tractor, the PTO driven machine stopped. This was really quite a nuisance. Many times the operator would like to have that machine continue running. After Cockshutt introduced the first continuous running power take-off in this country, other tractor manufacturers followed in close order; and two different types of PTO's appeared: the continuous and the independent. Their differences are still defined in standards of SAE and ASAE. Suffice to say, with a continuous running power take-off, the traction drive can be stopped and let the power take-off continue to run, but the power take-off cannot be stopped while the traction drive is running. The tractor has to be stopped to disengage the PTO. With the independent power take-off, there are independent clutches for both systems so either one is completely independent of the other. It was during this same period that the hydraulic lift pumps were taken from traction drive shaft and mounted on the engine or driven by the independent power take-off so that they would also continue to run even though the tractor master clutch was disengaged. Both of these items were very desirable from tractor operation point of view. It was also during this period that the application of hydraulics with the remote cylinders came into being. Deere was one of the pioneers in this effort.

In the late 1930's and during the 1940's, many implements, such as balers and forage harvesters, were being equipped with separate engine drives for the working mechanism, but were still being towed by tractors. The independent (and/or continuous running) PTO's on tractors resulted in successful PTO driven balers and forage harvesters. Separate engine drives disappeared except where self-propelled machines were developed.

During the 30's and 40's, tractors were getting transmissions with more and more gear ratios so that they could be better matched to the various jobs that they were required to do. It was about 1954, that the first partial range power shift transmission was introduced in the Farmall Super M-TA by International Harvester (Fig. 9). Minneapolis-Moline had a similar transmission about the same time. This permitted on-the-go shifting to a lower ratio for each of the major gears in the regular transmission. Today almost all tractors have on-the-go shifting either throughout the complete range of gears or at least through a significant number of gears. These transmissions result in energy conservation in actual agricultural service although it does not show up in the regular testing of tractors according to the SAE Tractor Test Code. The first tractor to have full range power shifting in all gears was the Ford Selec-O-Speed in about 1957. This provided on-the-go shifting in all gear ratios, even in reverse, without the use of a master clutch. Other sophisticated transmissions for special uses had been provided, such as Case-A-Matic which had a torque converter with a lock-up

Fig. 9 - Super M-TA, about 1954

in the mid-50's, and the International Harvester full hydrostatic infinitely variable transmission in the mid-60's. These transmissions had not received as wide acceptance as the partial range power shift transmissions.

Allis-Chalmers introduced the first agricultural tractor with a turbo-charged diesel engine in about 1962 (7). Turbo-charged engines had been used in construction equipment tractors prior to this, but this was the first specifically for agricultural use. Others followed in a very few years and the use of diesel engines became more and more popular. Gasoline and LPG declined rapidly during the late 60's. By 1970 a number of companies no longer provided carbureted engines in tractors. Turbo-charging is significant in the agricultural machinery field because it allows manufacturers to build engines of various power sizes with the same manufacturing tools. Turbo-charging an engine may also improve fuel efficiency.

The advent of factory-supplied roll over protection on agricultural tractors is credited to Deere and Company in about 1965 (Fig. 10). International Harvester and others followed in very close order. Since that period, many more items relating to operator environment and safety have been introduced at an ever accelerating rate. A large percentage of all tractors in the 90 horsepower and over class now have operator enclosures with special sound barriers, sound absorptions, air conditioning, radio, and so forth. No question, much has been done in the last ten years relative to making the tractor operator's life more desirable. We must acknowledge that the operator enclosures with air conditioning and filtered air were used in the self-propelled combine field before they were on tractors. This is understandable when one considers the environment during the harvest season; those operators were deserving of it.

Fig. 10 - Deere tractor with ROPS

Fig. 11 - International Harvester 3588, 1979

Four-wheel drive tractors have become more popular in the last 10 to 15 years even though there were some four-wheel drive tractors way back in the teens and even before. They did not receive wide acceptance until about the mid-60's. We have chosen to give considerable credit to Versatile Manufacturing Company as being the one that introduced the first articulated four-wheel drive tractor that was produced in significant volume for agricultural use. More recently, these four-wheel drive tractors, particularly those of the articulated type, have found their way into the row crop use (corn, cotton); and we may see more of this in the years to come (Fig. 11).

Has all this sophistication been just a desire to have something fancy? We think not. Let us examine this farm productivity graph (Fig. 12). In 1900, one farm worker supplied the material for approximately seven people. By 1930 he was supplying the material for about 9.8 or 10 people; by 1950, about 15-1/2 people; by 1960, about 25 or 26; 1970, about 47; and today, 58, 59, maybe 60. We're sure that this improvement in productivity is not all a result of just providing mechanical power for the farmer. His own inventiveness in how to produce a crop, better seed, better fertilizer; all such things enter into it. By and large, historians and economists believe mechanization has probably accounted for about 50% of the improvement. Even if it is less, it is still astounding. The wonderful thing about all of this is: as power has been applied to agriculture, its been done at an ever more efficient rate; that is, with more efficient use of basic energy (Fig. 13). In the 1910 to 1920 period, tractors produced about 1.25 to 1.4 kilowatt hours per each liter of fuel. This has improved until today's tractor produces about 3 PTO kilowatt hours per liter of fuel, about a 230% improvement. Data from drawbar tests in the conventional performance testing manner indicate about a 270% improvement. This drawbar data, however, does not take into account the gains that can be obtained through shifting on-the-go, less time at the ends, and so forth. So really, the improvement is more than indicated here. We think it is interesting to observe this next graph (Fig. 14) which shows the average power at the PTO or belt in a given year at Nebraska and then ten years later; ten years; ten years; and so on, up to the present. Note that in 1920 the power of the average tractor was 25 kilowatts. Then it rose a little bit, but during the 40's, the war years, as we might expect, there wasn't much gain. Most effort was being put into the war and not in developing new equipment. After the war, tractors were in big demand and manufacturers produced machinery of the pre-war design. Then power started to pick up again, and by 1979 the power of the average tractor tested at Nebraska was 104 kilowatts. These are just averages. They do not represent the extremes.

We are sure some important milestones have been missed; nevertheless, a pattern has been established by engineers of the past and engineers of today. Engineers have been an important part in adding to the quality of life throughout the world. The physical demands on the farmer have been reduced tremendously. The challenge is here for the engineer of tomorrow. We're sure he will accept the challenge and add to the glorious past.

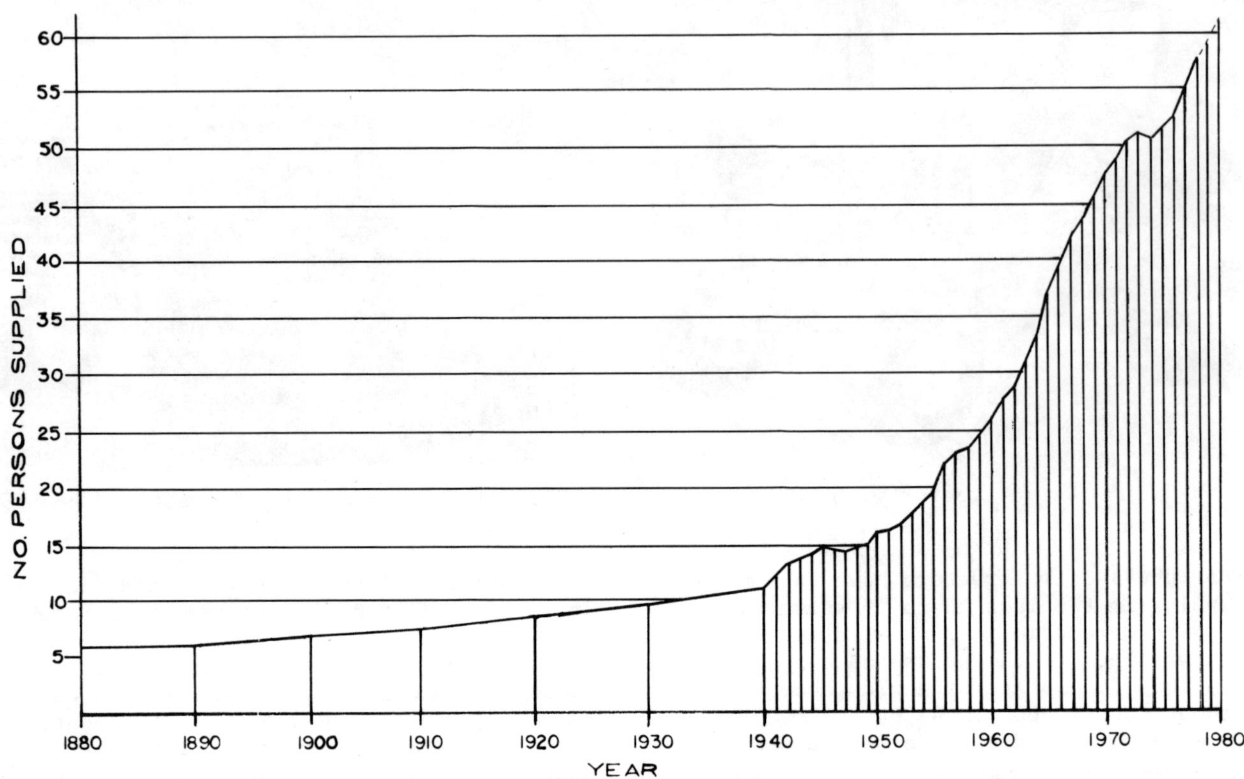

Fig. 12 - Farm productivity graph

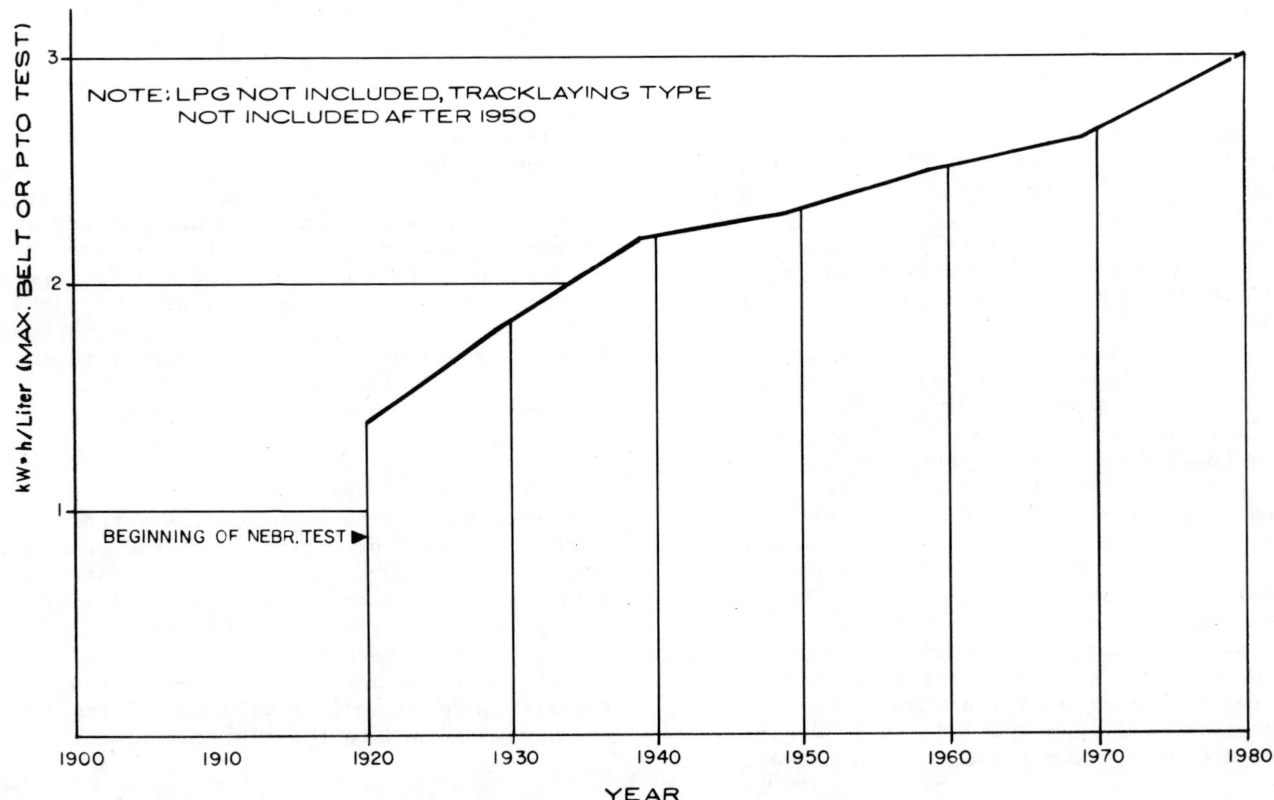

Fig. 13 - Tractor fuel economy graph

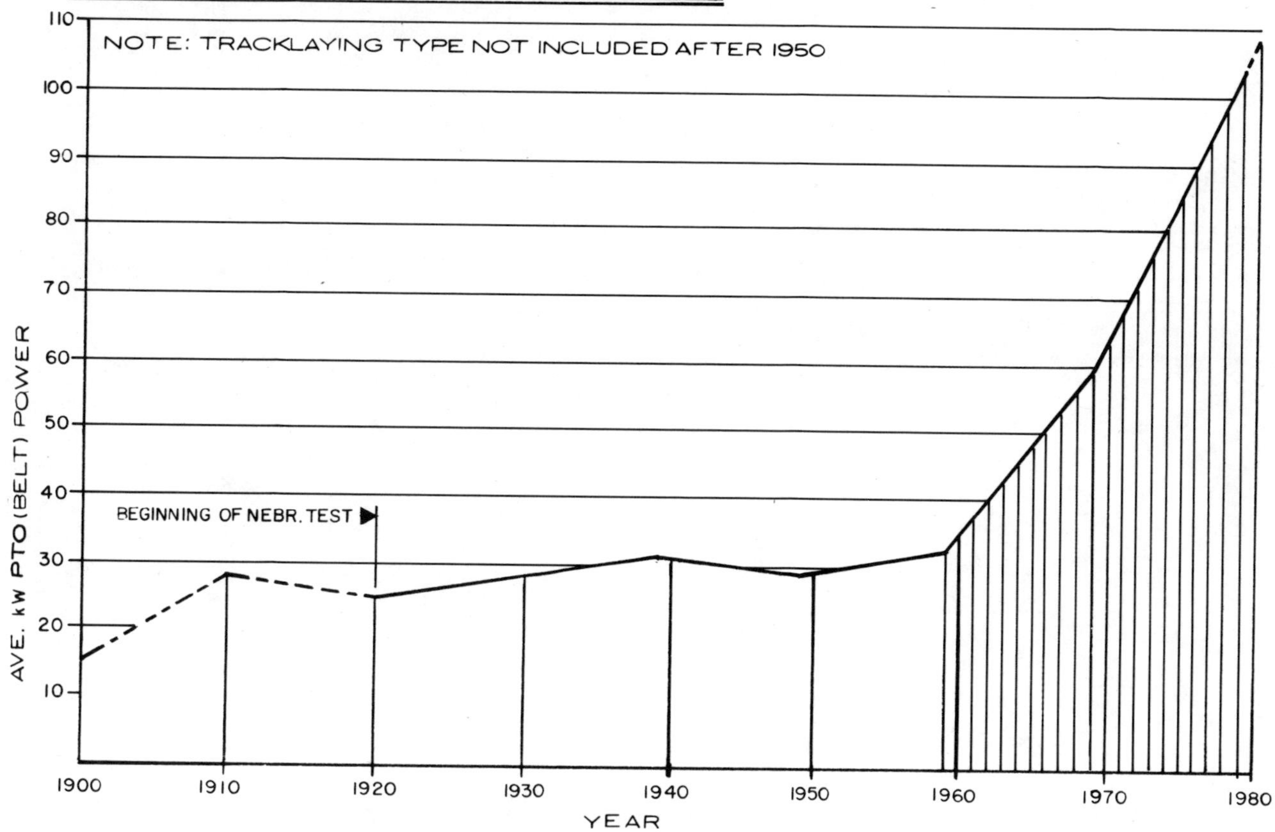

Fig. 14 - Average tractor power graph

REFERENCES

1. FIEI (Farm and Industrial Equipment Institute), "Men, Machines and Land." FIEI, 401 N. Michigan Avenue, Chicago, Illinois 60611, Booklet, 1974.

2. S. H. Rosenberg, "Rural America a Century Ago." American Society of Agricultural Engineers, 1976.

3. A. Stefferud, et al., "Power to Produce." The United States Government Printing Office, The United States Department of Agriculture, Washington, D.C., Yearbook of Agriculture, 1960.

4. R. B. Gray, "Development of the Agricultural Tractor in the United States." The American Society of Agricultural Engineers, Part I 1956, Part II 1958.

5. C. Frazer, "Tractor Pioneer - The Life of Harry Ferguson." Ohio University Press, 1973.

6. L. F. Larsen, et al., "Thirty Years of Nebraska Tractor Testing." American Society of Agricultural Engineers Technical Paper 76-1045, 1976.

7. C. W. Gray, "Notes on the Development of the Farmall Tractor." Unpublished, supplied by International Harvester Company, 1932.

8. Bureau of the Census, "Statistical Abstract of the United States." United States Department of Commerce, 1979.

9. Bureau of the Census, "Historical Statistics of the United States, Bi-Centennial Edition." United States Department of Commerce, 1976.

10. "Nebraska Tractor Test Summaries." University of Nebraska, Department of Agricultural Engineering, Lincoln, Nebraska, 1920, 1929, 1939, 1949, 1959, 1969, and 1979.

Society of Automotive Engineers, Inc.
400 COMMONWEALTH DRIVE, WARRENDALE, PA 15096

92 page booklet.

Printed in U.S.A.